Manufacturing Engineering Modular Series

Logistics & the Out-bound Supply Chain

John Meredith Smith

Penton Press

Publisher's note
Every possible effort has been made to ensure that the information contained in this book is accurate at the time of going to press, and the publishers cannot accept responsibility for any errors or omissions, however caused. All liability for loss, disappointment, negligence or other damage caused by the reliance of the information contained in this handbook, of in the event of bankruptcy or liquidation or cessation of trade of any company, individual; or firm mentioned, is hereby excluded.

Apart from any fair dealing for the purposes of research or private study, or criticism or review, as permitted under the Copyright, Designs and Patents Act, 1988, this publication may only be reproduced, stored or transmitted, in any form, or by any means, with the prior permission in writing of the publisher, or in the case of reprographic reproduction in accordance with the terms of licences issued by the Copyright Licensing Agency. Enquiries concerning reproduction outside those terms should be sent to the publishers at the undermentioned address.

First published in 2002 by

Penton Press
an imprint of Kogan Page Ltd
120 Pentonville Road
London N1 9JN
www.kogan-page.co.uk

© John Meredith Smith, 2002

British Library Cataloguing in Publication Data
A CIP record for this book is available from the British Library

ISBN 1 8571 8032 1

Typeset by Saxon Graphics Ltd, Derby
Printed and bound in Great Britain by Biddles Ltd, Guildford and King's Lynn
www.biddles.co.uk

Contents

Introduction   1

1. **Objectives and Measures of Performance**   3
   1.1 Introduction   3
   1.2 Performance of the distribution system   4
      1.2.1 Total delivery time   4
      1.2.2 Delivery reliability   5
      1.2.3 Order completeness   6
   1.3 Customer service level   7
   1.4 Cost   9
   1.5 Setting the strategy for distribution   9
   1.6 The task ahead   12
   1.7 Summary   15

2. **Order Management and Forecasting**   17
   2.1 Introduction   17
   2.2 Order processing   18
   2.3 Forecasting   22
      2.3.1 Judgemental forecasting   23
      2.3.2 Statistical forecasting   24
      2.3.3 Associative prediction   31
   2.4 Order processing systems   31
   2.5 Summary   36

3. **Stock Management**   38
   3.1 Introduction   38
   3.2 Order processing and stock records   40
   3.3 Stock handling and storage   42

	3.4	Accounting for inventory	45
	3.5	Stock control	48
		3.5.1 Control of primary stock	49
		3.5.2 The calculation of economic order quantity	52
		3.5.3 Control of safety stock	56
	3.6	The effects of multiple warehousing	59
	3.7	Summary	60
4.	**Transport**		**62**
	4.1	Introduction	62
	4.2	Types of transport	62
		4.2.1 Road	62
		4.2.2 Rail	63
		4.2.3 Sea and inland water	64
		4.2.4 Air	64
		4.2.5 Pipelines and cables	64
		4.2.6 Multimodal and intermodal	65
	4.3	Packaging	66
	4.4	The transport decision	68
	4.5	Responsibility and ownership	70
	4.6	Vehicle routing and scheduling	75
		4.6.1 Types of route	75
		4.6.2 Selecting a route	78
	4.7	Summary	83
5.	**Managing the Supply Chain**		**85**
	5.1	Introduction	85
	5.2	The supply chain and the demand chain	87
	5.3	Push and pull logistics	91
		5.3.1 Customer delivery expectation	91
		5.3.2 Cumulative lead time	92
		5.3.3 The fundamental principle of logistics	93
	5.4	Inventory and demand amplification	98
	5.5	Establishing a domestic distribution system	100
	5.6	Summary	105
6.	**International Distribution and e-Business**		**107**
	6.1	Introduction	107
	6.2	Value density and logistics reach	109
	6.3	Logistics reach and the market	110
	6.4	Distribution and inventory strategies	113

6.5	Facilities location	117
6.6	The impact of the Internet	118
6.7	Summary	120

Appendix: Typical Examination Questions — 122

Bibliography — 126

Background and Rationale of the Series — 127

Index — 129

Introduction

At the turn of the 19th and 20th centuries, manufacturing was still largely a craft industry in which the skill of the individual engineer or craftsman was the key competitive weapon. Most manufactured goods served domestic markets in the developed economies of Europe and North America, with raw materials provided from the European empires overseas. The early years of the 20th century saw the development of scientific management through the work of Frederick Taylor, Alfred Sloan and others, but World War II devastated much of the industrial infrastructure of Europe and depleted North America of its supplies of raw materials, so that in the immediate aftermath product availablity became the competitive parameter; any company that could get a product to market could sell it.

The second half of the 20th century witnessed the development of manufacturing technologies both in terms of production processes and systems of control, so that manufacturing today is highly automated, quality products roll out of factories and greater efficiencies mean lower prices. As we enter the new century, delivery performance has again become the critical success factor, whether through the speed of delivery, the precision of delivery, its reliability or flexibility. As consumers, we expect to be able to specify precisely which products we want, in what quantities, where we want them and when we want them, and have manufacturers respond to our needs within increasingly short time-frames. Logistics is becoming an important competitive factor in many manufacturing sectors.

This book sets out the basic principles of logistics and the management of the out-bound supply chain for engineers and manufacturers. It is aimed at undergraduates and postgraduates of manufacturing and engineering management, but will also be of interest to managers of businesses faced with the need to understand and improve their product distribution systems.

Chapter 1 introduces the issues faced and considers how to measure the performance of a distribution system. The next three chapters consider the three key elements of product distribution: order processing, including forecasting and information systems; stock management and control; and transport and packaging. The final chapters consider the dynamics of the supply chain, the design of domestic and international distribution systems, and explore the impact of the Internet and associated advances in information and computer technology.

1

Objectives and Measures of Performance

1.1 Introduction

When we talk of the supply chain, we refer to the network of processes, resources and enterprises that provides for the supply of goods from raw materials through processing and production to the end consumer. Logistics is concerned with the organization, co-ordination and control of the flow of goods through the supply chain. Most enterprises are concerned with managing both their *in-bound* and *out-bound* supply chains. The former comprise the flow of raw materials, components and products, which they purchase to manufacture and supply products to their customers and to support their business operations. The latter is the flow of their products from source through a distribution system to their end customers (Figure 1.1). The control and organization of the flow of goods in the in-bound supply chain is the concern of in-bound logistics and includes the procurement activity; the control and organization of the flow of goods within the company is the concern of internal logistics, the control and organization of the flow of goods in the out-bound supply chain is the concern of out-bound logistics.

In this book, we shall be concerned primarily with the out-bound supply chain and out-bound logistics; however, for companies trading business to business (that is to say not trading directly with end consumers), their out-bound supply chain is a part of their customer's in-bound supply chain and vice versa.

The out-bound supply chain and its associated logistics, or *product distribution system*, is concerned with the flow of goods from the point

4   Logistics and the out-bound supply chain

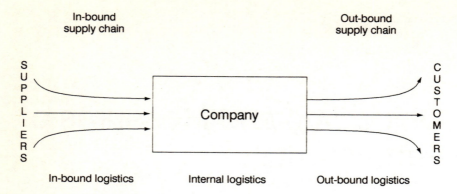

Figure 1.1 *The supply chain*

of production or manufacture to the point of consumption. Some products, for instance high-value capital goods such as aircraft, may go direct from the factory to the end user; others such as food and consumer disposables may pass through many hands on their way to the end consumer; furthermore, some products like water and electricity have their own distribution systems unique to the product.

The principle aims of product distribution are to get the product:

- to the right place;
- at the right time;
- in the right quantity;
- at the right quality;
- at the lowest cost.

1.2 Performance of the distribution system

The performance of a distribution system is measured by the customer in different ways.

1.2.1 Total delivery time

One measure is the total delivery time measured from when the customer places an order to when the product is received. In practice, this is made up of a number of components (Figure 1.2).

1. The time it takes to transmit the order from the customer to the supplier; this may be in part a function of the selling

process if contracts have to be agreed and signed, but for many products today electronic communication systems have brought this time down to the infinitesimally small.
2. The time it takes the supplier to process the order on receipt; this may involve checking customer credit worthiness, checking product availability, making delivery arrangements and, in the case of orders to be delivered to foreign countries, raising the appropriate export paperwork.
3. The time it takes to identify the items to be despatched in the supplier's stores, collect them together and pack and label them ready for delivery. Some, but not all, of these activities may be undertaken in parallel with those of (2) above.
4. The time it takes to physically transport the product from the supplier to the customer.

In many instances, where a product does not pass direct from its manufacturing source to the end consumer but passes through other hands, many of these processes are repeated at each stage of the product distribution process.

1.2.2 Delivery reliability

Another measure of performance of a distribution system is delivery time reliability. This is an indicator of whether the supplier delivers when they say they will. Although when a consumer buys goods from a retailer they expect immediate availability, for most other

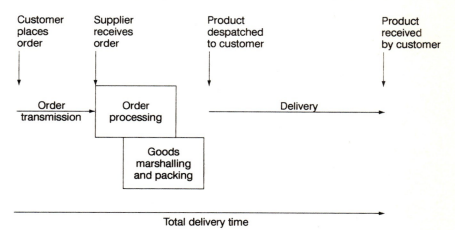

Figure 1.2 *Total delivery time*

products there is a delay between the placing of an order and the receipt of product. In some cases, the shortest delivery time is considered the best but frequently customers want and need to plan other activities in conjunction with the receipt of product from suppliers and having reliable delivery promises is important. Half a century ago, most factories planned deliveries on the basis of the month of delivery: the customer would be quoted a delivery date of, say, March. In practice, the customer had an expectation, not often met, that the product would be delivered in the first week at least, if not, 1st March; the factory, needless to say, felt it had met its obligations if the product was despatched on 31st March, arriving with the customer a day or so later. Progressively, delivery dates have been expressed more precisely as customers have become more demanding. Although week of delivery is still quoted in some instances, many markets now demand delivery dates set in days or even hours. A consumer ordering products on the Internet for delivery at home does not want to wait in all day for a delivery; shopping on the Internet is supposed to be a convenience and specifying and satisfying the time of delivery to an hour's precision is becoming a market necessity. On-time delivery, meeting delivery promises certainly to the day, more often to the hour or minute is an increasingly important measure of distribution system performance (Figure 1.3). Delivery reliability can be measured as the proportion of orders delivered within a given time-frame around the promised delivered date/time or by the variance or standard deviation of delivery time.

1.2.3 Order completeness

Orders frequently comprise more than a single item; this may be because the product ordered is made up of a single item or because the customer has ordered a quantity of a given product or a number of different products. In each case, the customer expects to receive the total order in a single delivery.

In the first instance, consider an order for a mobile telephone. So far as the customer is concerned this is probably a single product, but in practice they will expect to receive, in addition to the phone itself, a battery, a battery charger, connections from the battery charger to both the mains electricity supply and a car supply, earpiece and microphone, both with connectors to the main device, instructions manual, etc. If any one of these items is not available

Objectives and measures of performance 7

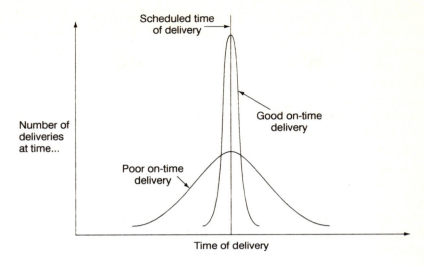

Figure 1.3 *On-time delivery performance*

the product cannot be shipped, or, if it is, the customer will be dissatisfied as the product may well be unworkable or at least less than satisfactory.

If a manufacturer orders a quantity of 300 items but only 250 are delivered, or if they order 20 different items in varying quantities but only 18 of them are delivered, then production schedules may be held up and the manufacturer is not satisfied with the supply.

Where an order is not complete and is not and/or cannot be delivered in its entirety, not only is the customer let down but if a second delivery has to be made the costs to the supplier are increased. Order completeness is another measure of the performance of a distribution system measured by the proportion of orders delivered complete.

1.3 Customer service level

The speed of delivery, the reliability of delivery and order completeness are simple measures of delivery performance often used by customers to assess their suppliers. Suppliers frequently set and claim to meet a customer service level often quoted as a percentage; however, on its own it is insufficient; being told that a 98% customer service level is offered tells one very little; 98% of what? To be of use, a customer service level needs to be defined as *the proportion of orders* of *given size* delivered within a given *time-frame*. A claim that 98% of orders of less than 100 items will be delivered within 24 hours is a meaningful measure and the customer can plan accordingly. It provides an indication of product *availability*.

Customer service may vary depending on the product (capital item, disposable, consumable, perishable) and the customer (consumer, distributor).

Delivery time can be enhanced by the use of fast transport such as air freight, couriers, etc., but, as we shall explore in Chapter 4, these tend to increase costs over standard surface freight services. Similarly, the deployment of inventory and the processing of orders physically close to the customer will reduce the total delivery time perceived by the customer, but holding stocks close to the customer, as we shall discover in Chapter 3, also increases the cost.

Delivery reliability can be improved by the use of a single and dedicated means of transport between source and customer. If goods are despatched from source to an intermediate point, where they are then unloaded and transferred to a different form of transport, there is a risk of delay and breakdown. The more intermediate points in the network of distribution, the less reliable any delivery may be. If products being delivered share transport facilities with other items, there is a risk of delay due to problems with the other items (Figure 1.4). Using public transport facilities and different forms of transport is often more cost-effective and the use of dedicated transport frequently incurs a significant cost penalty. As with delivery speed, the deployment of inventory and the processing of orders local to the customer will reduce the total delivery time perceived by the customer, but also at an increased cost.

Order completeness is more likely to be achieved by supplying product from large stocks and from having high-quality information systems providing accurate and timely data on availability and for stock planning purposes, all of which tend to add to cost.

The factors that lead to high perceived levels of customer service, delivery time and reliability and order completeness, are just those that simultaneously tend to increase costs. Likewise, those factors that may lead to reduced costs are those that mitigate against the measures improving customer service (Figure 1.5).

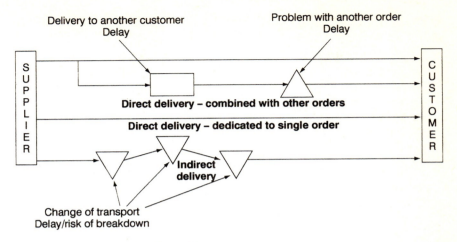

Figure 1.4 *Direct and indirect delivery*

1.4 Cost

To the distributor, cost is an important measure of performance and, to the extent that it is reflected in the price, the customer will also be a judge through this measure. The main cost drivers in a distribution system are the cost of order processing and order handling, the cost of inventory carrying and control and the cost of transport and packaging. The lower the costs, the less the outlay for the supplier and ultimately, probably the lower the price to the customer. However, as we shall see, most of the other performance measures of a product distribution system incur additional cost to enhance them so a careful balance has to be struck.

A whole series of trade-offs can be identified between the different factors and each has to be balanced to achieve the chosen objectives of the distribution system. It is the company's total business strategy that provides the framework for decision making in setting distribution objectives and, increasingly, it is the distribution function that has a key role to play in fulfilling the company's business objectives.

1.5 Setting the strategy for distribution

Michael Porter (1985) introduced the concept of competitive advantage as the cause of success for a company over its competitors.

Factors which lead to high levels of customer satisfaction	Factors which lead to lower costs
• Single mode of transport • Dedicated from source to customer • Air transport • Small quantity shipments • Individual consignments • Advanced order processing • Advanced Communications systems • Comprehensive Database systems • High supply chain visibility • Skilled order processing operators • Large inventory base • Dispersed inventory	• Use of bulk transport • Full return loads • Full containers • Low investment in systems • Low investment in communications • Low-cost employees • Few employees • Small inventory base • Centalized inventory

Figure 1.5 *Factors affecting performance and cost*

Competitive advantage can be achieved, according to Porter (1985), in one of two ways (Figure 1.6): either through cost leadership, by being the lowest cost provider of a product either over a broad target market or possibly over a narrowly focused target market; alternatively by unique product differentiation, again over either a broad or narrowly focused target market.

It could be argued that cost leadership is just another form of differentiation, but most other differentiators, whatever they may be, tend to increase cost whilst lowering cost tends to lead to lowest common denominators and hence a lack of any other differentiator. It is also to be questioned whether either the broad market but differentiated or the lowest cost but focused strategies are realistic. If lowest cost is the strategy, then it is most likely that this will be achieved by seeking economies of scale and hence high volume, so forcing the provider to a broad market strategy. If, on the other hand, offering a unique differentiator is the provider's strategy, then it is likely that the nature of the differentiation will only appeal to a limited market; the sharper the differentiator, the sharper the market focus is likely to be.

Whichever of the strategies, cost leadership or differentiation, the company selects to adopt, it is likely that logistics will have a part to play in their achievement. In the case of cost leadership, the consequential tendency towards a broad market approach necessitates distribution to that wide market. If the product is to be

Objectives and measures of performance 11

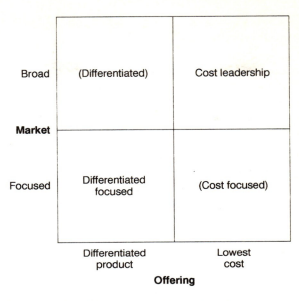

Figure 1.6 *Strategies for achieving competitive advantage*

sourced from a single factory then distribution to perhaps even global markets becomes a requirement; if the product is to be sourced from many factories, then the selection of their location and the patterns of distribution therefore become key logistics issues to be faced. In the case of targeted differentiation, then it is likely that some element of customer service will become a characteristic of the differentiation.

In achieving competitiveness, Terry Hill (1993) has distinguished between different classes of factor that influence a company's ability to succeed. Those factors that cause the company to beat the competition in obtaining orders are referred to as *order winners*; those for which the company has to meet a certain acceptable standard to be considered by the market as competitors are referred to as *order qualifiers*; other factors are less significant in creating competitiveness. In a differentiated product strategy, distribution is almost certainly going to be an order qualifier and is increasingly perceived in some markets as the order winner.

If a cost leadership strategy is to be adopted by the company, it may seek to find ways of minimizing distribution costs. Hence, ways of reducing stock-holdings and minimizing transport costs will be likely to take priority over actions to enhance the service level perceived by the customer, although there will probably be some order qualifiers to be met in this direction.

If a differentiated strategy is adopted, the fine tuning of the distribution system may be a force in creating the differentiation sought. Options may exist to provide high levels of product availability, very short delivery times, acceptance of small order quantities, or very precise delivery schedules as the company's unique characteristic in the market. The out-bound logistics system may provide a key component in the determination and achievement of a company's strategy.

Customer service is often a key contributor to customer retention and is frequently significantly less costly than winning new customers (Figure 1.7). Poor service in distribution as a result of incorrect or late delivery can lead to customer loss. Keeping customers content with an efficient and service-minded system is relatively simple compared with the job of winning over new customers from competitors who are also concerned with providing high levels of customer service.

1.6 The task ahead

In the next three chapters, we consider the main elements of distribution systems. Chapter 2 considers order processing and the life

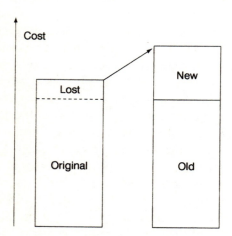

Figure 1.7 *Cost of customer loss and retention*

cycle of a customer order from the moment its arrives at the manufacturer until the customer has been satisfied. This includes:

- order receipt by the manufacturer;
- entry into the order book;
- allocation of stock or the initiation of manufacture;
- picking, packing and despatch;
- invoicing the customer.

We then go on to consider demand forecasting and the various techniques available, many of which are based on the information gathered and maintained by the order processing systems. These include:

- judgemental forecasting;
- statistical forecasting;
- analysis of time series;
- associative predictions.

Demand forecasting is an important factor in determining future courses of action for the business, in particular in stock control and distribution planning. The order processing system is also seen as an important source of information for monitoring the profitability and business performance of customers, markets and products.

Finally in Chapter 2, we look at how communication and information technology have influenced order processing systems in recent decades and how these developments have changed the distribution systems and are now impacting on a company's competitive strategies.

In Chapter 3, we consider stock management. First, we look at the administrative functions of stock record keeping, the allocation of stock to orders and the accounting for stock, all of which follow naturally from the order processing system. We then move on to look at stock handling and the processes of receiving and looking after stock in the warehouse. Then we investigate the various methods of stock control, aimed at achieving a balance between minimizing the costs of holding stock and the need to provide an adequate level of service to customers. Stock control systems are seen to consider stock as comprising:

- primary stock, which is used to satisfy demand between replenishment cycles;
- secondary or safety stock, which is held as a contingency against unpredicted variations in demand or supply.

We then consider the issues of managing stocks over a number of different warehouses and the effects on stock levels of holding stock in a multiplicity of locations. This provides us with the tools to take decisions about the size and shape of the stock-holding to meet business objectives.

Chapter 4 covers the carriage of goods and transport systems. It starts with a review of the different types of transport:

- road;
- rail;
- air;
- water;
- pipelines, cables, etc.

An essential factor in the transport of goods is packaging and this is the next issue studied in Chapter 4. We identify the three key purposes of packaging:

- product containment;
- product identification;
- product protection.

We then look at the transport decision; how to select the appropriate mode of transport and how much, if any, of the transport activity should be owned and controlled by the distributor and how much outsourced to public or private transport undertakings. This leads on to the issues of distributing goods for export and the different terms of pricing and responsibilities of the supplier and the customer under the standard terms of trade for exporting products.

Finally in Chapter 4, we look at the issues of routing and scheduling of vehicles making deliveries from a warehouse to a range of customers.

Chapter 5 considers the management and operation of the supply chain as a whole and its relation to the demand chain. We look at the distinction between push logistics and pull logistics and then establish the fundamental principle of logistics. This chapter then continues with the study of the interaction between the demand and supply chains and the problem of demand amplification, or the *bullwhip* effect. We then go on to explore different strategies for distribution systems and the alternative for locating inventory within the supply chain, channel strategy and the ownership and control of the sales activity.

In Chapter 6, we explore how the internationalization of business over recent decades is placing a greater emphasis on distribution systems and how this challenge can be met. It then reviews the options for international distribution and defines certain key measures for defining a framework from which international distribution strategies can be developed. A number of alternative strategies are then investigated that provide for different product characteristics and different customer service and market objectives. Some of the key factors affecting the selection of plant and warehouse location are considered.

Finally, the impact of the Internet and its effect on distribution systems is explored and the challenges it sets are identified.

1.7 Summary

The supply chain is the flow of goods from basic raw materials, through the stages of manufacture, to finished products and on to the end consumer.

For a company, the in-bound supply chain is the flow of purchased items for production, resale or consumption in their operating activities. The out-bound supply chain is the flow of finished products from production through a distribution network to end consumers.

The out-bound supply chain or product distribution system seeks to get the right products to the right place at the right time, in the right quantities at the right quality at the lowest cost.

Performance of the distribution system is perceived by the customer in terms of:

- total delivery time;
- delivery reliability;
- order completeness.

Customer service level is defined as the proportion of orders of a given size that are delivered within a given time-frame.

The cost of a distribution system is driven by:

- order processing and order handling;
- inventory carrying and control;
- transport and packaging.

A series of trade-offs can be found between these activities to meet different customer service objectives.

Companies may adopt cost leadership or focused differentiated strategies.

Distribution systems can contribute to cost leadership by low-cost systems, effective low-cost stock control and transport.

Distribution can contribute to differentiated strategies through high product availability, fast delivery, acceptance of small orders, and precise on-time delivery schedules.

Customer service contributes to customer retention, which is usually cheaper than going out to win new customers from the competition.

2

Order Management and Forecasting

2.1 Introduction

Order processing is the administrative process of receiving orders from customers, who may be either end consumers or companies operating as the next process in the distribution network, and ensuring that the right goods are delivered on time and to the right place. It frequently represents the interface between, on the one hand, the marketing and selling functions of a company, which are concerned with promoting the products and obtaining the orders, and, on the other hand, the distribution function, which is responsible for delivering the product to the customer. It can be seen as the first stage of the distribution process, and its main purpose is to record all customer orders received by the company to ensure their correct delivery to the customer and to see that a timely and accurate invoice is raised and despatched. The order processing function is also concerned with tracking the progress of the order from receipt to delivery and for providing any necessary status information, both internally and to the customer (Figure 2.1).

To the extent that it captures and holds data on customer orders received, the order processing system is frequently used to provide important information on patterns of both customer buying and product demand and usage for the purpose of forecasting and business planning.

Recent advances in information and communications technology have had a significant impact on order processing systems and these in turn are affecting the way we do business.

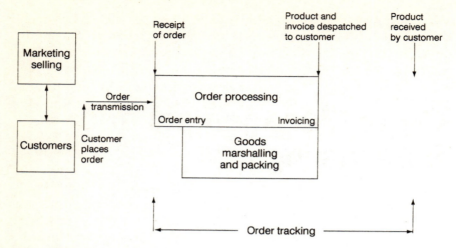

Figure 2.1 *Order processing*

2.2 Order processing

Taking orders sounds simple enough on the face of it, and in many instances, such as that of a market trader, it is. Typically, a market trader just takes an order, hands over the goods and receives cash in payment: the transaction is complete and the trader moves on to the next one. For the manufacturer, it is usually more complicated. The product may be more complex and so the order calls for careful specification and checking; the order may indeed include a large number of different items; the delivery may not be immediate, either because the customer does not want it yet, or because the product is not available; the customer may well be seeking credit, to pay against an invoice on receipt, rather than pay cash with the order; furthermore, not all items may be available and further manufacturing may have to take place prior to delivery. The goods then have to be packed and despatched or arrangements made for the customer to collect and take them away.

The first step on receipt of an order is *order entry*, recording the details of the order (Figure 2.2). This may first allocate a reference number to the order for internal identification tracking and recording the date the order was received. Then the customer's name and, frequently, an internal customer reference number are recorded, together with the customer's internal purchase order reference, if there is one, as this is likely to be the means by which the customer identifies the order. The address to which the goods are to be delivered and the address to which the invoice should be sent also need to be recorded at this stage. In many cases, if the order is from a regular customer, much of this information will already be held on the company's records and simply attached to the order record as it is entered. The next stage is to record details of the products ordered, and the quantities and the dates they are required. The order can now be priced and, where appropriate, assessed against the customer's credit worthiness. Product availability may also be assessed, from which it can be determined whether or not the order can be delivered to meet the customer's schedule requirements. If both credit worthiness and product availability are satisfactory, the order can be confirmed and appropriate notification of confirmation of the order communicated to the customer.

Should problems with credit worthiness be identified, these will probably have to be referred to sales and commercial functions of the organization, as these issues are normally beyond the authority of the distribution function. Should product availability not be sufficient to meet customer requirements then, in discussion with the customer, there may be some rescheduling necessary. Possible options include: rescheduling the whole order to a new delivery date; despatching some items to the customer's original schedule and rescheduling others to one or more new delivery dates; and the customer electing not to proceed with the order. Depending on the customer's decision, actions may then have to be initiated to obtain the product through manufacture and/or procurement.

Quite often, the customer does not specify a requirement date but simply seeks a delivery as soon as possible. In such cases, the assessment of stock availability and the possible need to manufacture or obtain the product from elsewhere leads to an estimated delivery date, which is offered to the customer. When all these issues have been resolved, the order can be confirmed.

Once the order is confirmed, a number of other actions have to be undertaken:

- stock allocation;
- manufacturing (if required);
- picking, marshalling and packing;

- arranging despatch and delivery;
- invoicing.

Stock allocation is frequently undertaken by the order processing function. However, as it is logically a function of stock management, it will be treated in Chapter 3, *Stock Management*. Likewise, arranging despatch and delivery may be undertaken by the order processing function but this will be treated in Chapter 4, *Transport*, where it fits more logically. Manufacturing is normally the responsibility of another function of the business, whilst picking, marshalling and packing are normally undertaken in the warehousing operation and will be considered in Chapter 3.

The instructions for manufacturing, if required, and for picking, marshalling and packing are, however, frequently generated by the order processing system. The progress of these activities is monitored by the order processing system so that order status is known and can be reported if necessary, both internally and to the customer (Figure 2.3). The picking instruction may be packed with the goods so that the customer can check what has been packed and a delivery note is frequently sent with the goods so that the customer can check the delivery.

Once the product is ready for despatch, the invoice can be raised and sent to the customer. In practice, in many situations the invoice may be raised in advance, but legally in most countries the invoice should be dated the day of despatch and any sales taxes calculated at the rates applying on that day. In the case of domestic despatches, the invoicing process follows naturally from the pricing of the order and completes the order processing activity.

In the case of despatches for export, there may be a great deal of additional documentation required, depending on the nature of the product, the means of transport and the regulatory regimes operating in both the exporting and importing countries. These documents include amongst others: a bill of lading; certificate of origin; a certificate of insurance; an export licence; and an import licence. In the case of goods being exported, there is also the issue of how they will be paid for and the raising of letters of credit, bills of exchange, etc. may also be called for. The detailed processing and documentation associated with exporting are beyond the scope of this book, but they can be significant and time-consuming activities that normally take place in parallel with the physical processes of distribution undertaken in the warehouse.

CUSTOMER ORDER

THE MANUFACTURING COMPANY		Order Number		
Customer No.		Date		
Customer Name		Invoice Address (if different)		
Address (for delivery)				
		Customer's Purchase Order		
Product No.	Description	Quantity Required	Date Required	Unit Price	Price

Figure 2.2 *Typical customer order*

22 Logistics and the out-bound supply chain

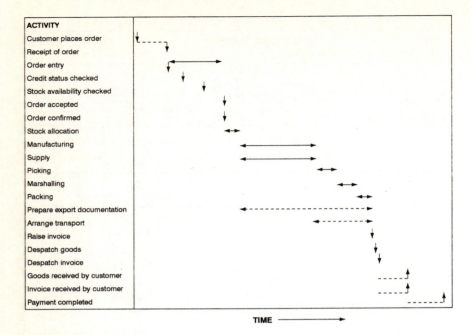

Figure 2.3 *Life cycle of an order*

2.3 Forecasting

Forecasts are required for a variety of purposes in running a business. To plan a business, we seek to predict future demand and then provide sufficient resource to satisfy that demand when it arises. Long-term forecasts are needed for product planning and development and for long-term capacity and resource provision; in the medium term forecasts are needed for material and stock planning, cash planning and budgeting and for manpower planning; in the shorter term forecasts are used for manufacturing, procurement and distribution scheduling. Many different techniques are used for the different purposes and ranges of forecast but they fall into three broad categories: judgemental forecasts; time series projections; and associative predictions. Where historical statistical data are available, often based on order processing systems, time series projections and associative predictions are feasible. Much long-term forecasting, in particular in respect of new technologies and in respect of variables for which no reliable statistical data are available, is more often based on the judgemental approach. All forms of forecasting merit consideration and many companies use combinations of different techniques to build their future plans.

2.3.1 Judgemental forecasting

This is based on knowledge and expert opinion of the environment, which is likely to affect the demand for the item to be forecast. This may include technological trends, economic assessment, competition, regulatory issues, etc. Some widely used judgemental forecasting methods are as follows.

1. *Expert opinion*. It is not usually difficult to find opinions on most issues and future product demand is no different! When forecasting demand for a given product or family of products, there are nevertheless a number of significant opinions that may well be relevant and able to identify important pointers. The sales force, for instance, is very likely to have a view on whether a product is about to do well or have some kind of a downturn; economists studying specific markets, particularly in respect of possible export sales, may well be able to provide insight into future trends in inflation, interest rates, consumer spending, etc.; technological experts may be able to give an indication on possible advances that might impact either product fabrication or use. Other experts may be able to throw some light on social trends, political developments and the potential for changes to the regulatory framework. These views and opinions are then consolidated to provide an informed forecast for the product's future. Such a forecast is normally in the form of trends rather than absolute values and should preferably incorporate upper and lower bounds of prediction and some estimate of possible error. Judgemental forecasting is particularly relevant for new and evolutionary products and for entering new markets. Questions such as assessing the likely impact of text messaging on postal services, the introduction of the Euro on the sale of newspapers, etc. are unlikely to be answered effectively by any other forecasting technique.
2. *Market surveys*. Another way of predicting the future sales demand is to ask the customers and potential customers. It is often relatively easy to conduct a survey of existing customers to ascertain what they are likely to buy in the short to medium

term. In businesses with a direct sales force, in fact, this is frequently an ongoing activity. A sound market survey, nevertheless, should not be restricted to existing customers as such a survey would effectively eliminate the possibilities of much market growth. The inclusion of a representative sample of non-current customers is therefore necessary and, to achieve this, the use of third party agents to conduct the survey is prudent, as buyers do not always reveal the truth to companies who they perceive as either suppliers or competitors to their suppliers. Some market research organizations maintain panels of consumers and customers selected specifically to provide representative samples of different markets. Separating the results of forecasts determined from, on the one hand, current customers and, on the other hand, non-customers frequently provides two different perceptions of the market and can lead respectively to different sales and marketing tactics, although these forecasts will need to be consolidated to provide production and distribution planning information.

3. *Delphi group forecasting*. This is a method of handling, in a structured way, information obtained from a range of expert opinion, or from a market research panel. The initial forecasts are all obtained quite independently of each other; they are consolidated and summarized to provide a first pass overall forecast, which is then passed back to the original sources for comment and, if appropriate, adjustment or revision. The revised forecasts are then consolidated in the same way and fed back again. This process is then repeated as many times as is necessary to obtain a consensus. The Delphi approach to forecasting has proved particularly valuable in respect of long-term forecasts involving technological change and its impact on consumer markets.

2.3.2 Statistical forecasting

This is based on the analysis of historical data for demand, and the order processing system is the primary source of such data. Time series forecasting is based on the extrapolation of past patterns of data to predict the future. In order to forecast future demand, it is most important to analyse past demand, if at all possible, rather than past sales, which may have been restricted by limited supply.

Unfortunately, not all order processing systems collect demand data, in fact almost all fail to capture unsatisfied demand data. The customer who walks into a shop and leaves without making a purchase, or who phones a company, makes a tentative enquiry but does not follow it up with any detail, does not leave a trail of demand data. They probably wanted something but it is not known what. *Orders received* data are frequently the nearest that can be obtained to demand data but any forecast made on the basis of these or similarly deficient data may need to be adjusted to allow for the difference between true demand and the data being analysed. Nevertheless, if orders received rather than sales despatched are used, this is a closer representation of actual demand.

Time series projections are based on a series of data recorded over successive time periods. Forecasting methods seek to analyse patterns in the time series in the past so as to predict its behaviour in the future. The pattern can be considered as comprising three or four components, which combine together to give a particular pattern. A long-term growth or decline in demand is referred to as the *trend* component of the time series and may follow various patterns, which may be modelled as linear, exponential, logarithmic, etc. (Figure 2.4).

Another detectable component of demand may be *cyclical* or *seasonal,* showing regular ups and downs in demand. Seasonal variations show up over annual cycles, whereas other cyclical variations

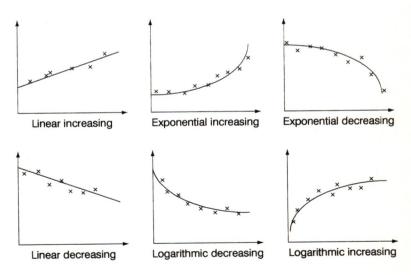

Figure 2.4 *Trends*

may be over longer periods such as economic cycles or product life cycles, or over shorter periods such as the month or week. Demands may also show a cyclical pattern *and* a trend or one cyclical pattern within another with or without a trend.

Finally, demand patterns frequently incorporate an *irregular* component, which is essentially random and so tends to be unpredictable and hence difficult to forecast.

In order to forecast future demand, it is necessary to analyse and understand the time series by trying to separate the various components of demand. The trend and cyclical components can potentially be modelled and hence are predictable; the irregular component by its nature is not so amenable to modelling. If the demand pattern is dominated by the irregular component, then it may be extremely difficult to make accurate forecasts using time series. If, however, the other components are dominant, then time series analysis becomes a viable method of forecasting.

Table 2.1 represents the quarterly demand for washing machines in thousands over a five year period.

Table 2.1 *Demand for washing machines*

Year	Quarter	Demand
1	1	3.6
1	2	3.2
1	3	4.8
1	4	5.0
2	1	4.2
2	2	4.0
2	3	5.1
2	4	5.5
3	1	5.6
3	2	5.2
3	3	6.0
3	4	6.3
4	1	6.2
4	2	5.8
4	3	6.3
4	4	6.9
5	1	6.8
5	2	6.5
5	3	7.1
5	4	7.5

In order to begin to determine the trend of the demand pattern, the quarterly moving average demand is calculated as follows. The first period for which four full quarters are available is the first full year, so the average is calculated as (3.6 + 3.2 + 4.8 + 5.0)/4 = 4.150; the next moving average is (3.2 + 4.8 + 5.0 + 4.2)/4 = 4.300, and so on; the first average calculated covers four quarters and so in time represents the demand between the end of the second and the start of the third quarter in the average; it is therefore listed against the third quarter (Table 2.2) and so on.

In order to determine the moving average that applies to actual quarters, the average of the moving averages at the start and the finish of each quarter is taken. The moving average for the third quarter of year 1 is therefore the average of 4.150 (the moving average at the start of the third quarter) and 4.300 (the moving average at the end of the third quarter) (4.150 + 4.300)/2 = 4.225; this is referred to as the *centred moving average* and that for each quarter is shown in Table 2.3, column 5 and plotted in Figure 2.5.

Table 2.2 *Moving averages*

Year	Quarter	Demand	Four quarter moving average
1	1	3.6	
1	2	3.2	
1	3	4.8	4.150
1	4	5.0	4.300
2	1	4.2	4.500
2	2	4.0	4.575
2	3	5.1	4.700
2	4	5.5	5.050
3	1	5.6	5.350
3	2	5.2	5.575
3	3	6.0	5.775
3	4	6.3	5.925
4	1	6.2	6.075
4	2	5.8	6.150
4	3	6.3	6.300
4	4	6.9	6.450
5	1	6.8	6.625
5	2	6.5	6.825
5	3	7.1	6.975
5	4	7.5	

Table 2.3 *Centred moving averages*

Year	Quarter	Demand	Four quarter moving average	Centred moving average	Seasonal irregularity factor
1	1	3.6			
1	2	3.2			
1	3	4.8	4.150	4.225	1.136
1	4	5.0	4.300	4.400	1.136
2	1	4.2	4.500	4.538	0.926
2	2	4.0	4.575	4.638	0.863
2	3	5.1	4.700	4.875	1.046
2	4	5.5	5.050	5.200	1.058
3	1	5.6	5.350	5.463	1.025
3	2	5.2	5.575	5.675	0.916
3	3	6.0	5.775	5.850	1.026
3	4	6.3	5.925	6.000	1.050
4	1	6.2	6.075	6.113	1.014
4	2	5.8	6.150	6.225	0.932
4	3	6.3	6.300	6.375	0.988
4	4	6.9	6.450	6.538	1.055
5	1	6.8	6.625	6.725	1.011
5	2	6.5	6.825	6.900	0.942
5	3	7.1	6.975		
5	4	7.5			

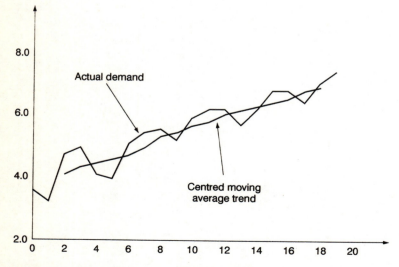

Figure 2.5 *Actual demand and centred moving averages*

If columns 3 and 5 of Table 2.3 are compared, it can be seen that whilst column 3 (the actual demand) exhibits seasonal peaks and troughs, column 5, because each value is a moving average that includes a value from each of the four quarters, has eliminated the seasonal effects and shows the general trend component of demand, excluding the seasonal component. By calculating the ratio of column 3 to column 5, the combined seasonal and irregular component of demand can be determined (column 6). Now the need is to separate the seasonal from the irregular components. To do this, the seasonal/irregular components for each quarter are taken and the average is determined (Table 2.4); by dividing the actual demand for the quarter by the seasonal factor for the quarter, the de-seasonalized demand (i.e. the trend plus the irregular component) is calculated (Table 2.5 column 8).

Having separated out the individual components of the demand pattern, each may be extrapolated to establish a forecast. The extrapolation may be done by a variety of methods. Trends may be extrapolated linearly using simple linear regression techniques. If the trend component can be represented by a straight line

$$v(t) = bt + c,$$

where $v(t)$ is the trend value at time t, b is the slope and c is the intercept on the v axis, then

$$b = \{\Sigma t x_t - (\Sigma t \Sigma x_t)/n\}/\{\Sigma t^2 - (\Sigma t)^2/n\}$$

and

$$c = \Sigma x_t/n - b\Sigma t/n,$$

from which further values of $v(t)$ can be calculated.

Basing trends solely on moving averages can be found to be giving too much weight in the forecast to the effects of past demand and insufficient weight to the more current levels. This can be overcome

Table 2.4 *Seasonal factors*

Quarter	Seasonal/irregular factors				Seasonal factor
1	0.926	1.025	1.014	1.011	0.994
2	0.863	0.916	0.932	0.942	0.913
3	1.136	1.046	1.026	0.988	1.049
4	1.136	1.058	1.050	1.055	1.075

Table 2.5 De-seasonalized demand

Year	Quarter	Demand	Four quarter moving average	Centred moving average	Seasonal irregularity factor	Seasonal factor	De-seasonal demand
1	1	3.6				0.994	3.62
1	2	3.2				0.913	3.50
1	3	4.8	4.150	4.225	1.136	1.049	4.58
1	4	5.0	4.300	4.400	1.136	1.075	4.65
2	1	4.2	4.500	4.538	0.926	0.994	4.23
2	2	4.0	4.575	4.638	0.863	0.913	4.38
2	3	5.1	4.700	4.875	1.046	1.049	4.86
2	4	5.5	5.050	5.200	1.058	1.075	5.12
3	1	5.6	5.350	5.463	1.025	0.994	5.63
3	2	5.2	5.575	5.675	0.916	0.913	5.69
3	3	6.0	5.775	5.850	1.026	1.049	5.72
3	4	6.3	5.925	6.000	1.050	1.075	5.86
4	1	6.2	6.075	6.113	1.014	0.994	6.24
4	2	5.8	6.150	6.225	0.932	0.913	6.35
4	3	6.3	6.300	6.375	0.988	1.049	6.01
4	4	6.9	6.450	6.538	1.055	1.075	6.42
5	1	6.8	6.625	6.725	1.011	0.994	6.84
5	2	6.5	6.825	6.900	0.942	0.913	7.12
5	3	7.1	6.975			1.049	6.77
5	4	7.5				1.075	6.98

by using the technique of exponential smoothing, which provides a means of balancing the past with the current, using the formula

$$F_{t+1} = \alpha v_t + (1 - \alpha)F_t,$$

where F_t, F_{t+1} are the forecast values at time t, $t+1$, v_t is the value of the series at time t, α is the smoothing constant ($0 \leq \alpha \leq 1$), and the actual value of α is determined by trial and error. A high value of α close to 1 will yield a forecast close to the actual for the last period; a lower value for α yields a forecast more heavily dampened by previous forecasts.

Cyclical components are normally treated as fixed and simply applied to the forecast trend as a repeated sequence of factors, although they may also in principle exhibit separate trends that can be analysed in the same way as above.

Irregular components are, by definition, irregular and therefore extrapolation is not strictly possible other than by assuming that the extent of irregularity will remain fixed. As long as the irregular component of demand remains small compared to the other two, its unpredictability is relatively important. When the irregular component becomes comparable with the others, forecasting becomes more difficult and unreliable.

2.3.3 Associative prediction

For some products, a forecast of future demand can be determined by measuring some other variable for which it has been shown that a close relationship exists between it and demand for the product in question. The form of the relationship needs first to be determined by monitoring actual demand against the expected related variable. For example, if short-term demand for a number of products has been shown to be weather related, by tracking rainfall or midday temperatures a forecast for a few days or a week may be determined using the sort of techniques described above for determination of time series trends.

This may provide data such as that shown in Table 2.6.

When the demand for BarBQ charcoal and the mean midday temperatures are plotted against the day, it can be seen that the demand follows the temperatures lagging by about four days (Figure 2.6).

Further analysis shows that the demand can be expressed as the relationship

$$D_d = 42(22T_{d-5} + 57),$$

where D_d is the demand on day d and T_{d-5} is the mean midday temperature on day $t-5$.

Other products and industries have been shown to either lead or lag the economy. Clearly, if one is in one of the latter then tracking economic indicators may give a useful means of forecasting future demand once the amount of lag has been established. Tracking the demand experienced by known leaders can equally provide reliable forecast information. In all cases, what is needed is an established relationship.

2.4 Order processing systems

Traditionally, order processing was co-located with the warehouse or stores; orders were received by post and order processing was

Table 2.6 *Demand for charcoal and midday temperatures*

Day	Demand for BarBQ charcoal	Mean midday temperature
1	15425	18
2	17959	17
3	17113	19
4	17597	22
5	18969	24
6	18074	26
7	20033	27
8	22784	27
9	24543	29
10	26361	29
11	27321	30
12	27194	28
13	29116	25
14	29185	24
15	30086	24
16	28177	29
17	25342	28
18	24523	29
19	24513	22
20	29135	25

undertaken by armies of clerks working with order books. Most of the effort was devoted to generating documentation for action: order confirmations, picking lists, packing lists, despatch notes, invoices, etc.; little time was left for forecasting and the provision of information.

Computer and telecommunications technology has obviated that need and over the past decades a number of technological developments have taken place to change the nature of order processing. These have both reduced costs and increased the speed and power of processing to the extent that order processing systems are becoming not just an essential administrative activity but a strategic competitive tool.

The modern computer-based order processing system dates from the 1970s and enables the order processing clerk to enter details of customer orders as they are received via a screen and keyboard, and is based on three or four basic files. A customer file maintains details of all customers including relevant addresses, credit status, pricing

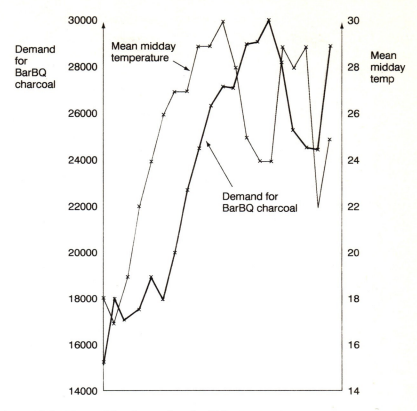

Figure 2.6 *Demand for charcoal and midday temperatures*

contracts, past consumption, etc. A product file maintains data on all products such as prices, costs, discount structures, consumption, etc.; this may be combined with, or separate from, a stock file, which maintains a record of current inventory levels including allocations, anticipated receipts from manufacture and/or suppliers and warehouse location data. Finally, an orders file maintains details of current customer orders from initial order entry through to despatch, invoicing and, in some cases, payment. The system undertakes all routine order processing tasks including order tracking, printing of action documents such as confirmations, despatch notes and invoices and maintains a historical record of consumption by both customers and products.

The development of the call centre and the associated telecommunications technology facilitates the centralization of the order

34 Logistics and the out-bound supply chain

processing function. All orders for a given geographical area, country or even globally can be handled in a single location, which is not necessarily co-located with any inventory, but which has complete access, via computer-based systems, to all the stock information necessary to process the order. This has the potential to reduce the overall number of operatives needed to process orders and can make all product information available to process all orders (Figure 2.7). The customer can be quite unaware of the location of the order processing activity, which may even be in a different country; from their perspective they have simply placed an order for delivery from a supplier company. What does need to be clear, however, is from what country and under what jurisdiction the trader is selling so that the law under which any contract is made is clearly understood by both parties.

Another technical development to have impacted on the order processing function in recent years is that of electronic data interchange or EDI. This is a means by which the computer systems of the

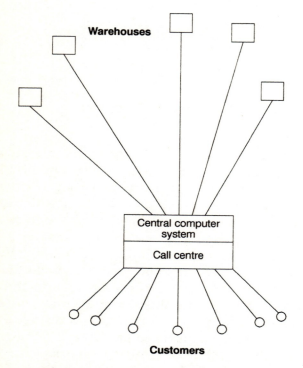

Figure 2.7 *Centralized order processing at a call centre*

buying organization can automatically order goods from the computer system of the selling organization without human intervention (Figure 2.8). EDI requires specialized software in each of the communicating computers and adherence to certain communications standards or protocols. Many large manufacturing companies have required their suppliers to install EDI facilities as a condition of remaining a supplier. Although there are potential savings from both the buying and the selling companies, the set-up investment can be significant and a deterrent to smaller companies. Due to the existence of a number of different EDI communications protocols, some suppliers can find themselves being asked to operate different systems for different customers with all the associated additional costs and effort.

More recently, the introduction of the Internet has facilitated much simplification of EDI from the user's perspective by using Internet software and communications disciplines to put EDI into effect. This is enabling many smaller companies to introduce EDI principles for placing and receiving orders and for the issue and payment of invoices.

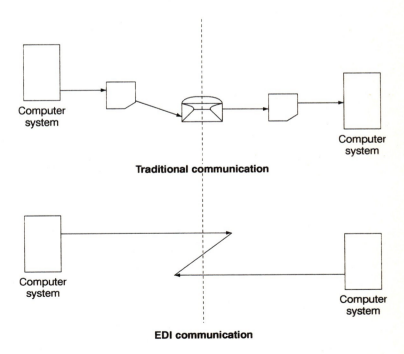

Figure 2.8 *Traditional and EDI systems*

These technological advances in order processing are providing significant benefits both to the manufacturer/supplier and to the customer/buyer. The facility to transmit orders from buyer to supplier virtually instantaneously provides the supplier with more time to respond to the buyer's needs within the time-span perceived by the buyer from placing an order to the receipt of goods. This is a major driver in motivating manufacturers to introduce order receipt by EDI, fax, e-mail and other electronic transmission means.

The same facilities enable orders to be received from throughout the world equally speedily and easily. Placing an order by EDI or e-mail is effectively independent of the location of the customer or the supplier. This facilitates increased market penetration by manufacturers and the opportunity to expand into foreign markets to an extent not otherwise possible.

Computer-based order processing systems make all product inventory potentially available to satisfy all customer orders. Orders, which might have been back-ordered for future delivery or possibly lost altogether, where orders were processed only against the local stocks, can be satisfied by identifying the location of stock, wherever it may be, which can satisfy the customer's needs. This not only provides better service to the customer but enables manufacturers and distributors to hold less stock. Particularly for those items that experience lower levels of demand, it is possible to centralize inventory holding or at least hold it at regional locations only. Manufacturers are able to significantly reduce their inventory holdings through the use of computer-based order processing systems without suffering any lowering of customer service levels.

Finally, computer-based order processing and associated order tracking systems enable distributors to monitor and control the progress and execution of orders throughout their life cycle more closely. This provides for greater delivery accuracy and timely availability of progress information to both manufacturer and customer; it also provides a means of access to order status during their progression and increased possibility of accepting changes from customers, hence increasing responsiveness and customer service.

2.5 Summary

Order processing is the interface between sales and marketing on the one hand, and distribution and manufacturing on the other.

Order processing comprises order entry, checking customer credit worthiness and stock availability, order confirmation, order progressing and reporting, stock allocation and sometimes invoicing.

It may also, in the case of export orders, include the preparation of specialized documentation needed for shipping goods abroad.

Forecasting may be undertaken by:

- judgemental methods including consulting expert opinion, conducting market surveys and using the Delphi method;
- statistical techniques of time series analysis, linear regression and exponential smoothing;
- associative predictions.

Order processing systems have benefited from advances in communications and information technology:

- computer systems and telecommunications have enabled the order processing function to be physically separated from the stock-holding location;
- call centres have enabled order processing to be centralized;
- EDI enables the buyer's and the seller's computers to talk to each other and to place and receive orders without human intervention;
- computer-based systems enable distributors to track and report on all stages of the order as it progresses from receipt to final delivery.

3

Stock Management

3.1 Introduction

Stock is held in a business primarily to provide availability of product when it is required. When we go to a shop for consumer goods, we expect to be able to leave with the items we were seeking to purchase. If the shop does not have the items, then the likelihood is that, in most cases, we will go somewhere else; only if we have strong loyalty to the shop or the brand is it likely that we will place an order for collection or delivery later. For industrial products, availability is just as important. Companies need from their suppliers the products and materials to keep their operations running, be it food for airline catering, computer supplies for a bank or components for a manufacturer; they all need them on time and hence available.

If product is produced and supplied continuously then, in theory, we can provide continuous availability; we only need to hold stock to cater for the emergency situation when production or supply is interrupted. For this reason, we do not normally hold stocks of electricity in our homes and only have a tank of water just in case there is a break in supply. However, if a product is not supplied continuously, which is the most frequent case, we need to hold stock to provide availability between replenishments.

Stock also serves some other purposes. The supply of product to satisfy a given need is frequently met through a chain of processes, starting with the acquisition of raw materials, their processing in some way their conveyance to a manufacturer, production, distribution, etc. (Figure 3.1).

Each of these processes is dependent upon its predecessor for a supply of work for it to continue in operation. By holding stock

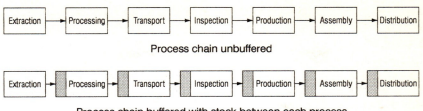

Figure 3.1 *Process chain unbuffered and buffered*

between each process in the chain, each is effectively buffered from its neighbours and can perform relatively independently. By using buffer stocks in this way, processes can be allowed to operate at a steady rate even if either supply or demand are fluctuating. By freezing fresh fruit and vegetables when they are in season and keeping stocks of frozen produce, we can have year-round supplies; by manufacturing Christmas fare throughout the year and holding stock, we can keep our production capacity operational throughout the year even though demand is highly seasonal. Both supply and demand tend to vary for a whole variety of reasons. If we were to operate without stock we would need to adjust supply to match demand and to do this we would need to have very accurate and far-reaching forecasts of demand. As we have already seen from the previous chapter, forecasting is not an exact science and there will nearly always be unpredicted variation one way or the other; another reason for carrying stock is to cater for these unexpected variations in both the supply of goods and the demand for them.

Stock is also held whilst it is being processed and so work-in-progress is also considered as stock. In addition to those items actually being worked on in a factory, there may well be items waiting to be worked on, waiting to be moved to the next stage of processing and also items being transported from one place to another. All are a part of the stock-holding of a company. Some products such as wine and cheese need to be stored for a period of time before they are sold to 'mature'; others such as medical supplies may have to be held whilst they undergo sterilization or some other process that makes them fit for use.

To summarize, stock can be said to be held for a number of purposes:

- to provide availability of product between replenishments;
- because supply is not continuous;

- as a buffer between processes in the supply chain;
- because it is in process or in transit;
- as contingency against unpredicted variability in demand and supply.

3.2 Order processing and stock records

In order to manage and control stocks, we need to maintain an up-to-date record of the quantities of stock held for each item in our warehouse. As we have already seen in Chapter 2, this is also a vital component of the order processing system. A basic stock record will hold, for each item held in the warehouse, a record of the quantity in stock and a history of recent receipt and issue transactions made against each item. In order to provide a full picture of what is happening to each item, it is common practice to separate into different transactions the different types of receipt and issue made. Most receipts in a warehouse will no doubt be from the manufacturing operation but some may come from outside suppliers or be returns from customers following an incorrect delivery, or returns from other in-house stock locations or from internal departments who have had some temporary use of the item in question. Such latter returns will normally have to have been cleared by quality assurance that they are in a satisfactory condition for use before being returned to stock. Issues, likewise, will normally be to customers as a sales transaction but may also be to other in-house stock locations or for internal use. As indicated in Chapter 2, it may be necessary to allocate stock to an order prior to despatch, in which case it is necessary to separate the transaction into two, one for allocation and one for despatch; the concept of *allocated* and *free stock* is introduced. This may be necessary for a variety of reasons.

Whilst many retailers hand over the goods as soon as they receive an order, others may take an order for delivery at a later date and it is necessary to ensure that once stock has been identified to satisfy that order, it is not used up for some other order or purpose within the company. In non-retail organizations, this is far more often the case; industrial customers may order a number of items in a variety of quantities to meet their operational needs and either supplier or customer may well want to consolidate these into a single delivery. Some products (for example, a mobile telephone) comprise a number of different items (handset, battery, battery charger, case, instruction manual, hands-free kit, etc.); if a retailer places an order with the manufacturer for 200 products, they cannot be delivered if the manufacturer is short of batteries, which probably come from an outside supplier. In this case, they will need to allocate 200 each of all the items that are available until the batteries come into stock, so as to ensure that they do not deplete their stocks of some other item below 200 before the batteries become available.

In this case, the situation may arise that some items of an order are available whilst others are not. Sometimes, partial delivery is possible but frequently it is not (e.g. you may order a personal computer with colour monitor and printer; the personal computer and printer may be available but not the monitor: in this case you would not accept a partial delivery. If, on the other hand, it was the printer that was unavailable you might be prepared to accept a partial delivery and take the printer later). A means of allocating or reserving stock to an order is frequently a necessary attribute of an order processing function. This is particularly relevant in capital goods and also in exporting, where it is frequently necessary to marshal the goods and then await the relevant export documentation before the goods can leave the warehouse.

When an order is taken, stock is allocated to it and the free stock (i.e. that stock which is still available to be allocated to orders) is reduced, but the quantity of physical stock remains unaltered. When the order is delivered, the total allocated quantity and the physical stocks are reduced but the free stock remains unaltered. Essentially, this facilitates the reservation of stock to orders taken; note that the free stock may be a negative quantity if the allocated stock exceeds the physical stock (Table 3.1).

When the order (line 2) is received, the order quantity of 100 is recorded as allocated stock, the free stock is reduced by 100 to 900 and the physical stock remains unchanged. When the order is ready for delivery (line 3), the allocated stock is reduced by 100, the free stock remains unaltered and the physical stock is reduced by 100. When the order for 400 (line 6) is received, the free stock is only 350 and so this order reduces it to a negative quantity, indicating a requirement for more stock, which is then met when the receipt transaction for 500 (line 7) is processed, increasing the free stock back to 450 and the physical stock to 1400.

At regular intervals, the stock record must be validated against the physical quantity held in the warehouse. This may be an annual or half-year event, frequently referred to as 'stock taking', in which all items in the warehouse are checked over a weekend or a few days during which the warehouse is operationally closed for business; or it may be undertaken on a product-by-product basis, checking so many products each week or each time a product has recorded a certain number of transactions. In this case, it is important to ensure that all items are checked at least once a year. After a stock check has taken place, it may be necessary to put through a stock adjustment transaction, which has the effect of changing the stock record up or down as necessary to reflect the physical quantity held. This will lead to a financial adjustment in the stock valuation.

3.3 Stock handling and storage

Typical warehousing operations are shown in Figure 3.2. Goods are received normally either from the manufacturing operation, or from an outside supplier from whence they normally are

Table 3.1 *Stock records*

Line	Transaction	Quantity stock	Allocated stock	Free stock	Physical
1	Opening stock		0	1000	1000
2	Order	100	100	900	1000
3	Delivery	100	0	900	900
4	Order	350	350	550	900
5	Order	200	550	350	900
6	Order	400	950	−50	900
7	Receipt	500	950	450	1400
8	Delivery	350	600	450	1050
9	Delivery	400	200	450	650
10	Order	300	500	150	650
11	Delivery	200	300	150	450
12	Delivery	300	0	150	150
13	Receipt	1000	0	1150	1150
14	Order	150	150	1000	1150
15	Order	250	400	750	1150
16	Order	200	600	550	1150
17	Order	400	1000	150	1150
18	Order	100	1100	50	1150
19	Delivery	150	950	50	1000
20	Delivery	250	700	50	750
21	Delivery	200	500	50	550
22	Delivery	100	400	50	450

checked for identity, quantity and quality; in the case of internal receipts from manufacturing this is usually dispensed with. The items then have to be located in the warehouse and put away. Location may be fixed depending on their identity or variable depending on available warehouse space, in which case the stock records will also have to record the location in the warehouse where the goods have been stored. It may be necessary to locate individual consignments of goods separately so as to be able to physically identify them by date or batch number, either because they are date-sensitive for use, or, for costing purposes, if items in different batches can have different costs. Fixed item location systems lead to ease of operation; warehouse personnel soon become familiar with locations and know where to locate each item. Variable location systems offer better space utilization, particularly where the quantities held of each item vary over time to meet variation in supply or demand; however, it is vital to have a reliable location recording system and there is always a risk of incorrect locating leading to items getting lost.

Customer orders are normally passed to the warehouse from the order processing activity once stock allocation has taken place, often in the form of a 'picking list'. This document, which replicates much of the information on the customer order, identifies the quantities of each item to be picked and may well also specify the warehouse location and the batch or date of products to be picked. It may be printed in location sequence to speed up the process of moving round the warehouse picking the items for the order.

Once all the necessary items for an order have been picked, they are marshalled in one place for packing and despatch and the despatch transaction is posted to the stock records.

In summary, the warehousing operation is concerned with the tasks of:

- receiving items from either suppliers or from own factory, recording their arrival and putting them away on the shelves or in the appropriate warehouse location;
- storing them safely and securely, preserving them in good condition and making them available as required;
- in response to demand from customers (or the next stage in the distribution chain) picking, packing and despatching the items to their appropriate destination.

44   *Logistics and the out-bound supply chain*

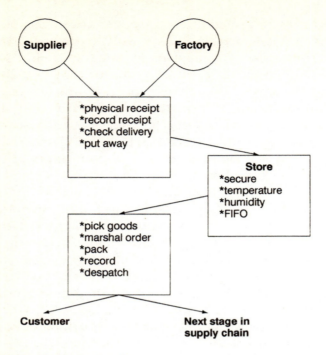

Figure 3.2 *Warehousing operations*

The greater the number of stages in the distribution process, the greater the number of times that the product is received, put away, stored, picked, packed and re-despatched. All of these activities add to the total material handling cost associated with the distribution of the product. Should this also involve a change in ownership then the further costs of invoicing and accounting are also incurred.

For many products, the only requirement of a warehouse is that it be secure, accessible to transport for deliveries and despatches, and reasonably dry. Other items such as fresh food, chemicals, pharmaceuticals, etc. have special requirements of temperature, humidity, etc., which can add significantly to the cost of storage. Stock should normally be held on a *first in first out* (FIFO) basis and the maintenance of accurate stock records is an essential part of warehouse management.

For all these reasons, holding inventory costs money. The cost of obtaining stock, whether purchased or manufactured, is incurred at the time of its acquisition; no income is received until it is sold; for this reason accountants often refer to it as a *deferred cost* or an *asset*. One of the most significant elements of holding cost is therefore the cost of the money tied up and not generating direct income. There is also the cost of the physical space that the stock occupies and the associated storage equipment such as racking and handling equipment. The manpower costs of the physical handling, stock recording and other related activities have also to be included. Other costs include insurance, shrinkage and opportunity costs. Shrinkage arises from stock losses not covered by insurance, including breakages, obsolescence, and theft; opportunity costs include the costs of lost business caused by holding the wrong stock.

Typically, these might range between:

money tied up:	5–25% pa of stock value
space:	1–5
handling:	1–5
shrinkage:	1–5
insurance:	1–4
opportunity:	1–6
	10–50%

It is unlikely that many business can carry stock for less than 10% of their unit value per year and it can easily exceed 50%.

Recently, the concept of the total cost of holding goods has been developed, which takes into account the cost to the business of carrying an item from the moment it is ordered from a supplier to the moment it is sold to a customer. As a result, an item that is held in stock for a month costs significantly more that an identical item only held by the company for a couple of days.

3.4 Accounting for inventory

Three of the most commonly used methods of accounting for inventory are:

- actual cost;
- average cost;
- standard cost.

The simplest is the *actual* cost, which is based on the purchase price of items as they arrive into a business, the costs being derived directly from the purchase accounting system. More sophisticated systems may incorporate a handling cost frequently calculated as a percentage of purchase price to include, in some way, the costs associated with such functions as purchasing, receiving, quality control, etc.

Whilst actual cost is conceptually simple, it is rather more complicated to use in practice because, at a time of varying prices, each order received may be at a different price and so each consignment has to be held and identified in stores separately. When an order is despatched, it is quite possible that identical goods are despatched that carry different costs. Stocks costed on an actual cost basis are frequently assumed to be held in stores on a FIFO basis for the purposes of inventory accounting. United States accounting requirements frequently call, nevertheless, for stock to be accounted for on a *last in first out*, or LIFO basis. This has the effect, particularly in inflationary times, of reducing the costs that can be inventoried.

In the example in Table 3.2, the opening stock of 1000 items are valued at £1 each so the initial stock value is £1000. The first three deliveries (lines 2, 3 and 4) are valued at £1 per item and the transaction values are simple to calculate. There is then a receipt (line 5) for which the unit cost is £1.10. The next delivery (line 6) can be completed with items valued at £1, but this uses up all the items at this cost, so the next deliveries (lines 7 and 8) use the £1.10 valued items. The next receipt (line 9) is valued at £1.20 per item, so when the delivery of 300 items (line 10) has to be made it comprises 100 items at £1.10 and 200 items at £1.20 = £3.50.

Table 3.2 *Stock accounting with actual costs*

Line	Transaction	Quantity	Unit value (£)	Transaction value	Stock quantity	Stock value
1	Opening stock		1.00		1000	1000
2	Delivery	100	1.00	100	900	900
3	Delivery	350	1.00	350	550	550
4	Delivery	400	1.00	400	150	150
5	Receipt	500	1.10	550	650	700
6	Delivery	150	1.00	150	500	550
7	Delivery	150	1.10	165	350	385
8	Delivery	250	1.10	275	100	110
9	Receipt	500	1.20	600	600	710
10	Delivery	300	*	350	300	360
11	Delivery	100	1.20	120	200	240
12	Receipt	200	1.10	220	400	460
13	Receipt	200	1.05	210	600	670
14	Delivery	500	**	565	100	105

After two receipts of 200 items each at values of £1.10 and £1.05 respectively, it is then necessary to make a delivery of 500 items. This can be made up, on a FIFO principle, of 200 items at £1.20 plus 200 items at £1.10 and 100 items at £1.05 = £565; or on a LIFO principle of 200 items at £1.05 plus 200 items at £1.10 and 100 items at £1.20 = £550.

Average costing methods avoid the problems of FIFO and LIFO by holding all like items at the same average cost, which is recalculated each time a consignment is received.

This also has its problems; the example in Table 3.3 shows the calculations. When the first new consignment of 500 is received (line 5), there are 150 items at £1.00 (= £150) and 500 items at £1.10 (= £550). As £150 + 550 = £700, the average value is £700/(150 + 500) = £1.08. When the next receipt comes in (line 9), we have 100 items in stock at £1.08 (although we really know they were received at £1.10) so the average is recalculated as £(500 × 1.20 + 100 × 1.08)/600 = £1.18. When we inspect the last two columns, we find that in line 6 we have 500 items in stock at a value of £538, which is £1.08 per unit, but by line 8 we have 100 items in stock at £106 or £1.06 per unit. The value should really be recalculated at each transaction, issues as well as receipts.

Table 3.3 *Stock accounting with average costs*

Line	Transaction	Quantity	Unit value (£)	Transaction value	Stock quantity	Stock value
1	Opening stock		1.00		1000	1000
2	Delivery	100	1.00	100	900	900
3	Delivery	350	1.00	350	550	550
4	Delivery	400	1.00	400	150	150
5	Receipt	500	1.10	550	650	700
6	Delivery	150	1.08	162	500	538
7	Delivery	150	1.08	162	350	376
8	Delivery	250	1.08	270	100	106
9	Receipt	500	1.20	600	600	706
10	Delivery	300	1.18	354	300	352
11	Delivery	100	1.18	118	200	234
12	Receipt	100	1.10	110	300	344
13	Receipt	200	1.05	210	500	554
14	Delivery	450	1.11	565	100	55

This method of costing is frequently used for ranges of goods bought from abroad in foreign currency and therefore liable to price fluctuation when converted to local currency. It does, however, mean that inventory values often change, calling for frequent stock re-evaluations and a consequential impact on stock provisions and reserves.

Standard costing systems set a cost for each item, which remains the same for a fixed period, normally a year, and is based on the predicted cost that will actually apply over the period. All transactions are calculated using that value and variances are calculated to assess performance against the standard. This system provides greater stability as the same cost is used for a period, frequently a year, but re-evaluations have to be made each time the standard is adjusted and then year-end adjustments have to be made to reflect the true costs and inventory values. Many manufacturing companies use standard costing systems, as they provide a means of monitoring performance in production by comparing actual to standard cost, and they eliminate the stock value variations caused by one of the other methods.

3.5 Stock control

Managing stock levels so as not to hold any more stock, and so not to incur any more cost than is necessary, but, at the same time maintain availability, is an important activity. Stock is held to meet demand between replenishment and also to cover against variability in demand and supply.

The stocks held to meet demand between replenishments is referred to as the *primary* or *cycle stock*; the stock held as contingency against variability is referred to as the *secondary* or *safety stock* (Figure 3.3).

The cycle stock (primary) includes the stock held to meet demand between replenishments (this could be zero if replenishment is continuous on a just-in-time basis). It also includes the work-in-process stock and the delivery or *in transit* stock. The cycle stock may also include some anticipation stock, which is that stock bought into play to meet *anticipated* demand over and above the normal, or to allow for an anticipated shortfall in future delivery.

The safety stock (secondary) includes stock that is held as cover against possible variability of demand, possible variability of supply and also variability of process trigger.

Stock management 49

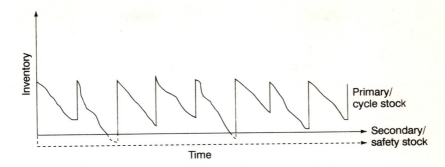

Figure 3.3 *Primary and secondary stock*

3.5.1 Control of primary stock

Primary stock is there to provide availability between replenishments.

Replenishment may be either at a fixed delivery frequency (Figures 3.4 and 3.5) or to a fixed order quantity (Figure 3.6). When replenishing at a fixed frequency of deliveries, the objective is to

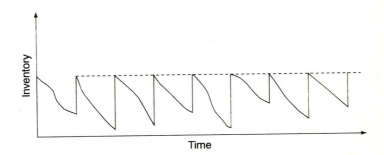

Figure 3.4 *Fixed delivery frequency fixed stock level*

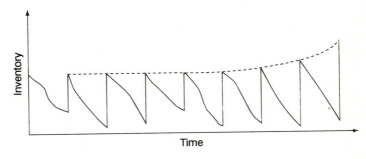

Figure 3.5 *Fixed delivery frequency increasing stock level*

order a quantity that will provide sufficient stock until the next delivery; this may be set at a fixed level, sometimes called an *imprest level*, or it may be a variable level depending upon circumstances and forecast demand. So, for example, in the BarBQ charcoal example in Chapter 2, the stock level sought on a daily basis would vary according to the predicted demand determined from monitoring the mean midday temperatures. Before each order is made, the stock level is assessed and the order quantity calculated to bring the inventory level back to that which is required.

The major supermarkets use this system to replenish stock in their shops; because they control the warehouses as well as the shops they are able to make deliveries very quickly so that the total replenishment time from order to delivery is short (often overnight). As goods pass through the checkouts in the shops, in addition to generating the bill for the customer, the shop's stock record is decremented and this information is automatically passed through to the warehouse, where the appropriate delivery is made up and despatched to the shop.

This system works effectively for goods for which regular and frequent deliveries are made because the replenishment cycle is short and the goods are fast moving. It does not provide a means of minimizing either stock-holding or delivery costs.

In a scenario of relatively steady demand, if the frequency of delivery is increased the order quantity will decrease and the average level of stock will also decrease (Figure 3.7).

Replenishing to a fixed order quantity is more often used when demand is a little less steady and the order quantity is determined to minimize overall costs. The question of when to order is determined from a knowledge of the average usage, the lead time and the safety stock level (Figure 3.8).

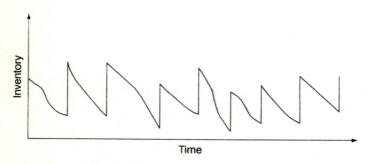

Figure 3.6 *Fixed order quantity variable frequency*

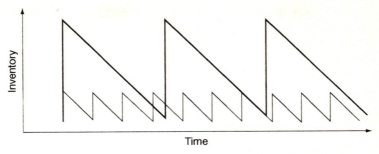

Figure 3.7 *Effect of varying frequency of delivery on stock level*

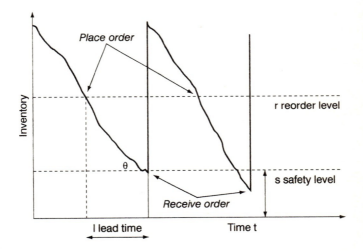

Figure 3.8 *Reorder point*

The objective is to place an order at such a time that under normal circumstances the inventory level will have just fallen to the safety stock level at the moment when the replenishment delivery arrives. The stock level at which the order needs to be placed can therefore be calculated as that when it is equal to the sum of the safety stock and the mean usage over the lead time.

So, if

u is the usage,

l is the lead time,

52 Logistics and the out-bound supply chain

and

s is the safety stock level,

then we can calculate the reorder level r as

$$r = ul + s,$$

or $r = l \tan \theta + s.$

3.5.2 The calculation of economic order quantity

As has already been noted, holding inventory costs money; the cost relating to each transaction results from the order placing process and the physical handling of each delivery in respect of receipt, inspection, putting away, etc., and its recording.

Effective stock management amounts to a balance between the costs of carrying inventory and the costs of handling the associated order and delivery transactions (Figure 3.9).

If

u is usage per unit time,

c is the unit cost of ordering/delivery,

$100i$ is the stock-holding percentage cost per annum

and

p is item unit cost,

then, if

order quantity $= x,$

then

average stock $= x/2,$

so, the average stock-holding cost is

$$V_1 = xip/2;$$

also, the interval between orders is

$$t = x/u,$$

and hence the number of deliveries in unit time is

$$u/x,$$

so, the ordering cost is

$$V_2 = cu/x$$

and the total cost is

$$V_0 = V_1 + V_2 = xip/2 + cu/x,$$

which has a minimum when

$$ip/2 = cu/x^2$$

and

$$x = \sqrt{(2cu/ip)} = Q_0,$$

where Q_0 is the *economic order quantity* or EOQ; sometimes referred to as the *economic batch quantity* or EBQ.

The minimum total is

$$V_0 = \sqrt{(2cuip)}$$

and the optimum delivery frequency is

$$F_0 = u/Q_0 = \sqrt{(uip/2c)}.$$

The EOQ calculated in this way (known as Wilson's formula), whilst having a number of shortcomings, remains the basis for much stock management in terms of balancing the replenishment cost with the stock-holding cost.

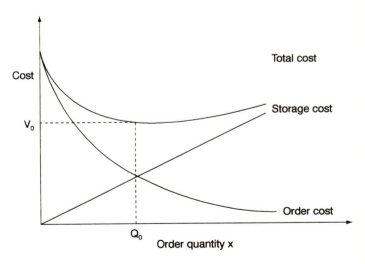

Figure 3.9 *Economic order quantity*

Usage is not normally a constant and so an average value has to be assumed, which may have more or less validity. The replenishment ordering and delivery cost may also (and normally does) include a volume-related cost element in addition to a fixed one (although this does not radically change the calculation, unless it is significant in comparison with the fixed cost).

The effect of discounts

The unit cost may vary with the order quantity if discounts are available for larger quantities; this can be incorporated in the calculation to determine whether or not to take the discount (Figure 3.10).

Using the same notation as before, let

$$p_1 = \text{unit cost when } x < a,$$
$$p_2 = \text{unit cost when } a \leq x < b,$$
$$p_3 = \text{unit cost when } x \geq b,$$

then

$$Q_1 = \sqrt{(2cu/ip_1)},$$
$$Q_2 = \sqrt{(2cu/ip_2)},$$
$$Q_3 = \sqrt{(2cu/ip_3)},$$

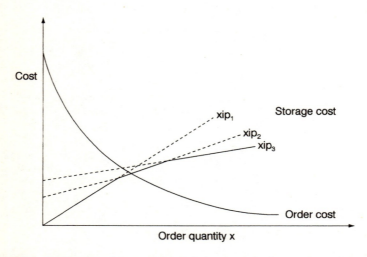

Figure 3.10 *Economic order quantity with discounts for larger quantities*

but since

$$p_1 > p_2 > p_3,$$
$$Q_1 < Q_2 < Q_3.$$

Should they all fall in the same price band, as they quite often do, only one quantity applies; if they fall in different price bands, the higher band will normally deliver the lower cost. If Q is just below a price break, then the cost can be investigated at the break point. For example, suppose usage is

$$u = 5000 \text{ pa},$$

stock-holding cost is 10%

$$i = 0.1,$$

order/delivery cost is

$$c = 9,$$

unit cost is

$$p_1 = 100 \text{ if } Q \leq 100,$$

unit cost is

$$p_2 = 95 \text{ if } Q > 100,$$

unit cost is

$$p_3 = 90 \text{ if } Q > 200,$$
$$Q_1 = \sqrt{9000} = 94.9, \text{ say } 95,$$
$$V_1 = \sqrt{900000} = 949,$$

if order quantity >100 and so $p_2 = 95$, then

$$Q_2 = \sqrt{(9000/9.5)} = 97.33, \text{ which is } <100,$$

if order quantity $= 100$

$$V_{stock} = 100 \times 95 \times 0.1/2 = 475,$$
$$V_{order/del} = 5000 \times 9/100 = 450,$$
$$\text{Total } V = 925 < 948,$$

so take the discount and order 100.

Stock replenished through production

The economic order quantity calculation can also be modified to take into account internal supply when the replenishment takes place not instantaneously but over a period of time (Figure 3.11).

Using the same notation as before, but introducing

$$\text{production rate } r \text{ per unit time,}$$
$$\text{set up cost } s,$$

then

$$\text{if batch size} = x,$$
$$\text{interval between orders is } T = x/u,$$
$$\text{No. of deliveries} = u/x \text{ per unit time,}$$
$$\text{period of production } t = x/r,$$
$$Q_o = \sqrt{\{2(c + s)u/[ip(1 - u/r)]\}},$$
$$V_o = \sqrt{\{2uip(c + s)(1 - u/r)\}} = ip(1 - u/r)Q_o,$$
$$F_o = u/Q_o = \sqrt{\{uip(1 - u/r)/2(c + s)\}}.$$

3.5.3 Control of safety stock

Safety stock is that stock held as contingency against variability in demand, in supply, and in the way the order is triggered. Where volumes are sufficiently large, demand variability may be modelled by the normal distribution.

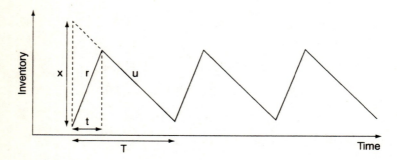

Figure 3.11 *EOQ when replenishment is over a production period*

Assuming that demand variation follows a normal distribution, it can be expected that 68% of demand falls within 1 standard deviation of the mean value of demand, 95% within 2 standard deviations and 99% within 3 standard deviations. However, safety stock is needed only for those variations in demand that are in excess of the mean; for those variations where demand is less than the mean there will be an expectation, to the contrary of excess stocks, so no safety stock is needed (Figure 3.12).

Where safety stocks are required, it follows that:

- safety stocks covering 1 standard deviation will leave 16% of orders unsatisfied;
- safety stocks covering 2 standard deviations will leave 2.5% of orders unsatisfied;
- safety stocks covering 3 standard deviations will leave 0.5% of orders unsatisfied;

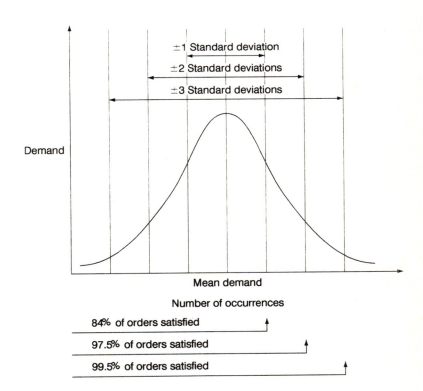

Figure 3.12 *Effect of variation in demand*

or:

- safety stocks covering 1 standard deviation will provide 84% service level;
- safety stocks covering 2 standard deviations will provide 97.5% service level;
- safety stocks covering 3 standard deviations will provide 99.5% service level; and
- safety stocks covering 1.64 standard deviations will provide 95% service level.

In general, the safety stock necessary to meet demand variations can be expressed as:

$$\text{safety stock, } s_d = Z\sigma\sqrt{l},$$

where

Z = the standard normal variate,

s_d = the standard deviation of demand,

l = the lead time.

In the same way, if variation in supply performance is normally distributed, similar patterns are found and the safety stock needed to meet these variations can be calculated by

$$\text{safety stock, } s_l = Z\sigma d,$$

where

Z = the standard normal variate,

s_l = the standard deviation of lead time,

d = demand volume per unit time.

Why should the variation in either demand or in supply performance be normally distributed? In practice, they are frequently not. In fact, many variations in supply show a much greater propensity to lateness than early delivery, and demand variations can also exhibit skewness to one side or the other. In addition, the time at which a replenishment order is placed can also exhibit patterns that lead to supply difficulties and the need for safety stock. This is the problem of the *order trigger* (Figure 3.13).

The *replenishment trigger* is frequently generated by a stock movement causing stocks to fall to a critical level, e.g. order point. If order quantities are small relative to total demand, the usage causes a steady decline in stock; if, however, the order quantities are large relative to usage, this is not so and the order may not be triggered until stock has fallen way below the order point (see Figure 3.13). This has the effect of apparently reducing the lead time.

Research on the setting of safety stocks continues, the arguments go on raging and most of the statistical theories can be shown to have practical deficiencies. However, the formula

$$s_d = Z\sigma\sqrt{l}$$

has been shown empirically to be valid in many circumstances where the lead time variation is small and it is still a valid and worthwhile start to calculating a safety stock. Trial and error can then be used to tune safety stocks upwards or downwards as necessary.

3.6 The effects of multiple warehousing

There are many instances where stock, instead of being held centrally, is dispersed to many locations to facilitate greater proximity to the consumer. For a given volume of total demand, the primary stock needed will not change by splitting the stock up into different locations; on the other hand, the amount of safety stock required will increase with increased number of stocking points.

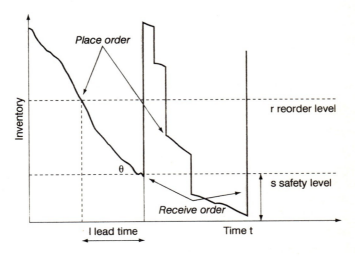

Figure 3.13 *Effect of variation of order trigger*

Assuming in this situation that each of the stock-holding points is identical in size and experiences the same pattern of demand, it can be shown that the safety stock increases in proportion to √(number of stocking locations).

As the number of stocking points increases, changes in transport costs may well offset stock-holding costs; these will be discussed further in Chapter 4.

The values of reorder point, order quantity and safety stock calculated as above are based on the demand and supply patterns at a single stocking point. If there are multiple stock-holding points, then the primary stock-holding for the total will be the sum of the primary stock-holding for all the stock-holding locations in the system. However, this is not true for the safety stock, which increases with the number of warehouses according to

$$s = S\sqrt{n},$$

where

S = the safety stock for all the stock in one location,

s = the equivalent safety stock spread over n locations.

This formula assumes that the stock levels, supply and demand patterns are the same at all locations, which is clearly unlikely to be true in the majority of cases; however, the basic principle of increasing safety stocks holds true.

A balance usually has to be found between a large number of stock-holding locations offering proximity to customers and a smaller number carrying less inventory but increasing transport times and costs. Different products call for differing levels of service and so different configurations. These will be explored more thoroughly in Chapter 5.

3.7 Summary

Stock is held to provide availability of product. It also is needed:

- because supply is frequently not continuous;
- as a buffer between processes;
- because it is in process or in transit;
- as a contingency against unpredicted variability in demand and supply.

Stock records maintain a record of:

- physical stock;
- free stock;
- allocated stock.

Free stock may be negative.
  Warehouse operations handle:

- stock receipts, including checking and putting away;
- storage, including security and protection of goods in the appropriate environment;
- maintenance of first in first out, dates and batches;
- picking, marshalling, packing and despatch.

Stock can be accounted for by:

- actual cost (as manufactured or purchased);
- average cost;
- standard cost.

Primary stock provides for availability between replenishment.
  Replenishment may be based on fixed delivery frequency to a fixed or variable stock level or imprest.
  It may also be based on a fixed order quantity.

$$\text{The reorder level} = \text{usage} \times \text{lead time} + \text{safety stock}.$$

$$EOQ = \sqrt{[2(\text{ordering cost} \times \text{usage}) \div (\text{unit holding cost})]}.$$

EOQ may be adjusted to allow for:

- quantity discounts;
- replenishment from production over time.

Safety stock provides for contingency against:

- variability of demand;
- variability of supply;
- variability of order trigger.

Setting safety stock $= Z\sigma\sqrt{l}$ is a useful starting guide.
  When stock is held in more than one location, the level of primary stock is unchanged; the level of safety stock increases with $\sqrt{(\text{number of locations})}$.

4

Transport

4.1 Introduction

The main objective of transport in the distribution system is simply to move products from their source to the customer. In meeting this objective, however, the questions that have to be answered are numerous. Which type of transport is to be used: road, rail, sea, air, or some dedicated form of product delivery such as a pipeline? How many different modes of transport will be employed in getting the product from source to the customer? How many breaks should there be in the journey and where should they be? What route should be taken from source to customer and what is the best time to undertake the journey? To what extent should the transport activity be subcontracted to transport service organizations and to what extent should it be undertaken as an in-house activity? Who should take responsibility for the transport of products: the supplier or the customer? There is no single answer to all these questions. In this chapter, we shall try to throw some light on the issues and explore the advantages and disadvantages of different solutions.

4.2 Types of transport

4.2.1 Road

Road transport is capable of providing a door to door service without any break in the journey to change from one vehicle to another; so it can, unlike any other means of transport, move a product from anywhere to anywhere else. Road vehicles, being relatively small, can also be transported by other means of transport such as ships and aircraft, hence providing the ability to offer direct delivery in the same vehicle even when other means of transport become a necessity. In most countries, the road system is perceived by the authorities as being a part of the national infrastructure and is funded out of taxation, effectively providing a subsidy to road transport. As a result, it can frequently offer competitive prices relative to other forms of transport, an advantage that is enhanced further by the fact that the road haulage industry comprises many independent operators fighting for their share of the market, making it highly competitive. Road transport is also, for most companies, the only mode of transport for which an in-house operation is a realistic option.

Although there has been significant investment in road infrastructure in most developed countries over the past decades, this has been matched by a corresponding increase in road traffic, leading to serious congestion in many cities and urban areas. Estimating delivery times and maintaining schedules becomes difficult with resultant risks to meeting on-time delivery targets.

4.2.2 Rail

The rail networks have experienced major investments and developments in certain European countries, in particular France and Germany where they are significantly subsidized by the taxpayer, whilst other developed countries by contrast have invested relatively little. Rail has the capacity to transport both heavy and high volumes of goods over land at relatively modest costs and at a higher travelling speed than road. Bulk transport of fluids and solid aggregates can be undertaken by rail using dedicated rolling stock where appropriate. In comparison with other modes of transport, the rail systems tend to maintain their schedules relatively well, but overall journey times are frequently long due to time spent at the start and destination of the journey, where a change of transport mode normally takes place, and also where an interchange is required by the system. Although speeds are relatively high when the goods are moving, they tend to spend much time waiting. The relative inconvenience of rail compared to road has caused a relative decline in its use as an overland mode of transport in recent decades.

4.2.3 Sea and inland water

Ships are the oldest form of bulk transport and still provide the most cost-effective means of moving large quantities, whether measured in terms of weight or volume over large distances. If no land route (including tunnels and bridges) is available, water remains the sole alternative to air for most products and is capable of handling virtually all cargoes. Like rail, it can suffer from delays at the start and finish of the water sector and at interchanges; it is nevertheless reliable and relatively cheap.

4.2.4 Air

By contrast, air transport offers a means of rapid transport and continues to develop its extent and capability. It provides by far the quickest mode of moving goods over large distances, but compared to surface transport is expensive. The capacity of aircraft is also limited by technology and air is unable to compete with rail or water-borne transport for the movement of bulk loads. Although not as flexible as road in providing door to door transportation, the use of light aircraft in countries such as Australia enables air transport to offer a higher degree of flexibility of source and destination than rail or water transport.

4.2.5 Pipelines and cables

Gas, water, oil, electricity, etc. are products that lend themselves to the use of dedicated transport systems. Such systems involve massive initial investment and need high volumes of movement to justify them. Once installed, however, they provide very cost-effective mechanisms for the transport of their product and relieve other systems of possible congestion. This type of distribution also exploits the fact that the product will flow; unlike most other forms of transport it is unidirectional. This has both advantages and disadvantages. If the customer wishes to return the product, another mode of transport has to be sought. However, there being no potential capacity for return journeys, there is no potential waste if this is not used. There is a wide choice available to the distributor in deciding how to convey the product to the customer, all having particular advantages and disadvantages relative to different products and markets.

A comparison of these modes of transport is shown in Table 4.1.

Table 4.1 *Comparison of modes of transport*

	Road	Rail	Water	Air	Dedicated (cable/pipeline)
Cost	Medium	Relatively low	Low	High	Low (after high capital outlay)
Speed	Medium	High	Low	High	High
Reliability	Poor in cities	Relatively high	High	High	Very high
Capacity	Low	High	High	Low	Very high
Transportable by others	Yes	No	No	Not normally	No
Impact on environment	Perceived as high	Relatively low	Low	Relatively high	Variable
Direct source-destination	Yes	Not normally	Not normally	Not normally	Yes
Government subsidies	Yes	In some countries	Not normally	Not normally	No

4.2.6 Multimodal and intermodal

In some instances, to avoid the need to change mode of transport, one type carries another. Probably the most common these days is the vehicle ferry upon which road vehicles are carried on a boat, effectively enabling a single vehicle to transport goods from source to destination without unloading and reloading the goods (although the vehicles themselves have to undergo this process at the ferry terminals). Road vehicles are also carried by trains, such as Eurotunnel, and also by some aircraft; some ferries can carry trains or at least railway rolling stock. Also, these techniques simplify the transport process, in many cases avoiding what otherwise might become an intermodal journey, which is one involving two or more means of transport.

When intermodal transport becomes a necessity, the critical issue becomes the means of transfer from one mode to another. Manual transfer is obviously time-consuming and costly and also subject to the risk of errors being made and the goods being damaged. This transfer process is significantly simplified today by the use of containers, which are essentially rectangular metal boxes looking rather like a trailer or a railway goods wagon without wheels. In order to effect simple intermodal transfer, the containers conform to standard sizes, which facilitates both lifting and transport. Standard containers are normally 8 feet by 8.5 feet by 20 or 40 feet in length. For measuring transport capacity, the 20 foot length container has become the unit of measure or TEU, the twenty foot equivalent unit.

4.3 Packaging

When transporting goods from source to customer, packaging is an essential feature of the product and the form this takes is often considered a part of the distribution process; it serves three basic functions:

- *containment* – to keep the product together, essential in the case of fluids but also necessary in many other types of product;
- *identification* – not only a description of the item but also reference numbers, production batch numbers, serial numbers, sell-by and use-by dates;
- *protection* – to protect the product from damage during the distribution and storage processes and also to protect from and identify unauthorized tampering with the contents of the product.

Obviously, many products like liquids, gases and powders need to be held in a vessel or box, simply to prevent them from dispersing or flowing away. There is also a need to put products into some form of container so as to know how much product there is. Hence, bottles, boxes, cans, etc. come in defined sizes, 1 pint, 1 litre, 1 gallon, etc. Others are designed to hold a specified weight of a given product such as 1 kg of sugar or a tonne of cement. Goods are also packaged for display purposes in shops, where the customer expects to be able to see the quantity of products on offer, such as the number of sweets in a bag or the colour and style of a shirt in a package. When fresh vegetables are offered for sale in a supermarket, many are pre-packaged in pre-defined packets of a given weight or volume that have already been priced. Even when unpackaged produce is purchased, the customer is frequently expected to package it so that it can be weighed and priced before passing through the point of sale terminal.

This leads naturally to the second purpose, that of identification. Although on the surface the nature of the product may be thought to be obvious, in practice this is often not the case. Even with a product such as milk, it may not be entirely clear by visual inspection whether it is full cream, semi-skimmed or skimmed; this information has to be provided on the label. Product identification also incorporates an element of product specification, i.e. what materials go to make up the product; particularly important for food and pharmaceutical products. Other product information that is increasingly needed on the label includes its place of origin, its date of manufacture and the production batch number or item serial number, so that traceability to a particular production process becomes possible in the event that this becomes necessary. For many products, the date beyond which they should not be displayed for sale, and the date beyond which use is not recommended is also provided in the identification. Increasingly, identification is effected by the use of bar codes, which facilitate automatic identification of the goods, not only at the point of sale terminal familiar in the supermarkets, but also on receipt and despatch by warehouses and when being loaded and unloaded onto different vehicles and modes of transport.

Finally, protection is the vital role of packaging whilst the goods are in the distribution process. Products need to packed in such a way as to avoid the risk of damage whilst in transit or storage; many products, such as foods and medical supplies, need to be sealed by the package to preserve freshness or sterility and may also need to be protected from possible contamination or tampering.

Packaging of many products frequently comprises two layers referred to as *primary* and *secondary* (or interior and exterior, or sometimes the consumer and industrial). The *primary packaging* is the box, bottle, package or vessel closest to the product, which contains the product and protects it from unauthorized access and preserves it until required for use. It frequently identifies the product and does so in a way readily recognizable to the customer. The *primary packaging* is often seen by the marketing function as a tool to promote both product and brand awareness. Coca Cola, Marmite and Quaker Oats are products readily recognized by their familiar and unique packaging. *Secondary packaging* is more concerned with protection of the product during the distribution process and is more the concern of the logistics function. Both primary and secondary packaging have to incorporate identification information. The primary packaging provides information for the consumer or customer, including a full or partial specification if that is what is required; the secondary provides information for both carriers and customs and regulatory authorities to enable them to ascertain contents and destination.

The needs of packaging over recent years has led to significant growth in the quantity being used. Globalization of products has resulted in more transportation, greater potential risk of damage and hence more protective packaging. Equally, globalization has created the requirement for identification information in many different languages; hence either the package has been increased in size to provide extra space, or products destined for different countries have had to have different packaging, creating the needs for more stock management of packaging materials.

Many packaging materials, both secondary and primary, are both bulky and combustible; this creates problems for both the supplier who has to store the materials, often in a special warehouse to meet fire regulations, and for the customer who has to dispose of the packaging on receipt. The disposal issue is leading to environmental concerns and to legislation requiring suppliers to facilitate its collection and either recycling or satisfactory ecologically acceptable disposal.

4.4 The transport decision

The cost of transporting products is a function of their physical characteristics (weight, volume, robustness, and physical state, i.e. solid, aggregate, liquid, gas, etc.), the distance to be travelled and the quantity of product to be moved. As a general rule, the greater the distance travelled the greater the cost (there is frequently a linear relationship between cost and distance for a given mode of transport); the greater the volume or weight, whichever is the driver of cost, the less the unit cost of transport. When transporting goods from source to destination by any non-unidirectional mode of transport, consideration has to be given to the inherent cost of returning the vehicle from the destination back to the source. If a return load can be found, the cost of the return journey can be borne by the return load; if not, then the original load may have to bear the cost of the return journey, so increasing the cost. Using a third party carrier to undertake a part of the journey provides the opportunity for that third party to find load for the return journey, so reducing cost. Hence, the transport strategy employed by many manufacturers is to move goods in bulk from source (or as near as possible to the source) to a distribution point (or points) from which a number of destinations can be reached; transport from the distribution point to the final destination is then effected by another mode or means of transport. This transport of bulk goods, normally undertaken by a transport undertaking (railway, shipping, etc.) is referred to as *trunk* or *primary transport*. The transport from the distribution points to the final destination is undertaken for individual customer consignments and is known as *secondary transport* (Figure 4.1).

For the majority of consumer goods, the final stage of transport is actually undertaken by the customers themselves from the distribution point (or, in this case, retailer) to their homes. Small retailers do not order in sufficient quantities to facilitate economic bulk transport from many manufacturers to the retail premises, so the

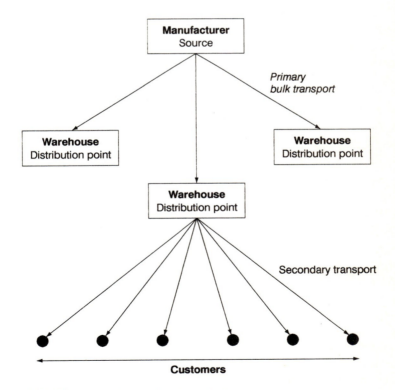

Figure 4.1 *Primary and secondary transport*

trunking is undertaken in two stages. In the first stage, the goods are taken from the manufacturer to a concentration point (wholesaler); on this leg of the journey the loads are bulked by the manufacturer. The second stage of the bulk transport is from the concentration point to the retailer, where the loads comprise goods from a variety of manufacturers but are bulked according to the retail destination. Large retailers tend to operate both stages themselves, operating their own bulk warehouses and effectively acting as their own wholesalers (Figure 4.2). Increasingly, major retailers have taken ownership of the physical distribution system and effectively manage the process from the manufacturer's gate through to their retail outlets.

For industrial products, there may be even more stages or *echelons* in the distribution process. Manufacturing plants may send their products to a regional warehouse, where products from global plants are marshalled for distribution to national warehouses and then onwards to local warehouses sited to optimize service levels to concentrations of customers. Such systems for industrial products are more often managed and controlled by the manufacturers themselves through dealerships and networks of agents and distributors (Figure 4.3).

Wherever the mode of transport changes, due, for instance, to the use of rail, sea or air freight, the change gives rise to further complexity. At each point, the goods have to be unloaded, checked and reloaded onto their next stage of transport, incurring both time delay and additional cost. The goods may also be held in inventory awaiting other items for the ongoing consignment, so adding further to cost. If this can be avoided and the goods transferred direct from the unloading/receiving process to the loading/despatching process, such inventory costs are eliminated; this is referred to as *cross docking*.

4.5 Responsibility and ownership

If, as individuals, we order some goods from a supplier, we assume that the supplier will arrange their delivery to us, frequently in their own vehicle. However, this is far from always being the case.

Within domestic markets, on receipt of an order from a customer most manufacturers take responsibility for arranging delivery to the delivery address specified by the customer. As it is the manufacturer

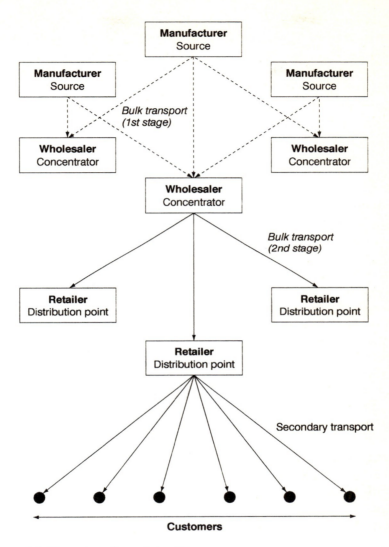

Figure 4.2 *Transport via a wholesaler*

(or supplier) who knows when the goods are ready for despatch, they are best placed to fulfil this role. The cost of delivery may well, in this case, be included in the price. In many cases, however, the cost of delivery is quoted separately and may vary depending on the delivery destination. The supplier, although raising an invoice at the point of despatch, normally takes responsibility for the product whilst delivering it and the customer only accepts responsibility on

72   Logistics and the out-bound supply chain

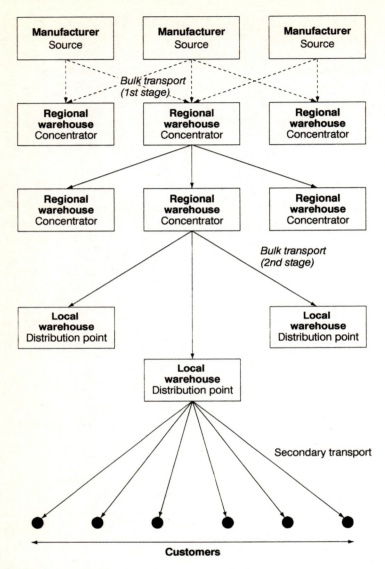

Figure 4.3 *Transport via concentrators and distributors*

receipt. Some companies, however, including both some manufacturers and major retailers are finding it cost-effective to organize the delivery of goods from their suppliers. By placing tight delivery (or, more accurately, availability) schedules on their supplier, they schedule vehicles to collect from a series of suppliers and combine the transport costs of a number of different consignments of goods all destined for their own use. In this case, the customer takes responsibility for the goods from the time of collection and may negotiate lower prices for so doing.

In international markets, this process is more formalized through the International Chamber of Commerce and the Incoterms standard terms of contract. A choice of 13 different conditions of trade are defined by the Incoterms 2000 (see Table 4.2); these range from *EXW (ex-works)*, under which the supplier only makes the goods available at the plant with it being the responsibility of the customer to arrange for their collection and carriage to wherever they are required with the customer accepting full responsibility for the goods from the point of collection, to *DDP (delivered duty paid)*, under which the supplier delivers the goods to a location specified by the customer with all duty and charges paid. Other commonly used terms of trade include: *FAS*, in which the goods are delivered at the supplier's expense to the quay alongside a vessel at a named port of destination, after which the responsibility for carriage becomes that of the customer (e.g. FAS Southampton Incoterms 2000); *FOB*, in which the goods are delivered at the supplier's expense on board a vessel at a named port of destination, after which the responsibility for carriage becomes that of the customer; *CIF*, in which the supplier bears the cost, including the costs of freight and insurance to a named port of destination, after which the responsibility for carriage becomes that of the customer.

The terms fall into four groups:

- *E-terms*, of which only EXW exists under which the seller only makes the goods available to the buyer on the seller's own premises;
- *F-terms*, under which the seller undertakes to deliver the goods to a carrier commissioned by the buyer;
- *C-terms*, under which the seller has to arrange carriage but without accepting certain risks due to events occurring after shipment;
- *D-terms*, whereby the seller accepts all costs and risks to a specified place of destination.

Note that FAS, FOB, CFR, CIF, DES and DEQ only apply to goods being transported by sea, whereas other terms can apply to any mode of transport.

Table 4.2 *Incoterms*

Term	Description
EXW	Price ex-works, not cleared for export, from a named place
FCA	Price, cleared for export, delivered to a carrier at a named place
FAS	Price, cleared for export, delivered to alongside a vessel at a named port of shipment
FOB	Price, cleared for export, delivered to on board a vessel at a named port of shipment
CFR	Price, cleared for export, including cost of freight, not insurance, to a named port of destination
CIF	Price, cleared for export, including cost of insurance and freight to a named port of destination
CPT	Price, cleared for export, delivered to a named place (excluding insurance cost)
CIP	Price, cleared for export, delivered to a named place (including insurance cost)
DAF	Price, cleared for export, and delivered to but not through a named frontier
DES	Price, cleared for export, and delivered, but not unloaded or cleared through customs, at a named port
DEQ	Price, cleared for export, delivered and unloaded from a ship, but not cleared of customs duty, at a named port
DDU	Price, cleared for export, delivered and unloaded from a ship, but not cleared of customs duty, to a named place
DDP	Price, cleared for export, delivered and unloaded from a ship, and cleared of duty, to a named place

Where goods are conveyed by rail, by sea/water, or by air, the transport function is normally subcontracted to an independent transport undertaking (the carrier). Where goods are transported by pipeline or cable, it is more often that the pipeline or cable is operated by the supplier, although in the United Kingdom we have recently established the operation of both the national electricity grid and the national gas pipeline system separately from both the suppliers and sellers of those products. In the case of road transport, the choice still remains whether to use a transport undertaking or one's own transport to convey goods. For primary bulk transport, the use of a transport undertaking means that one only pays for the distance the goods are transported, the problem of return loads rests with the haulier; however, if the journey is such that there is little chance of being able to find a return load or if the goods need specialized vehicles that mitigate against finding suitable return load, it is most likely that the haulier will pass the full cost of the return journey on to their customer and no saving is made. The use of the manufacturer's own fleet of vehicles offers both an advertising opportunity by displaying the company's name and image on the vehicles and, in the case of secondary transport, the opportunity for direct contact between the manufacturer's delivery personnel and their customers.

4.6 Vehicle routing and scheduling

Two significant issues arise for the distributor or the transport operator:

- how to allocate journeys or deliveries to vehicles;
- how to allocate customers to warehouses.

Different types of transport service may call for different approaches to route planning and scheduling. So-called *link-oriented* routes are used, where all locations along a path between two nodes must be served; postal deliveries, bus operations and waste collection services fall into this category. On the other hand, *node-oriented* routes involve connecting a number of nodes or delivery points in the most efficient way.

4.6.1 Types of route

Three different types of route can be identified as follows.

Radial routes follow a path along main radial roads from the depot, making successive deliveries or collections; finally, the driver makes a return journey on which the vehicle may be either empty or full, depending on whether it has been delivering or collecting. Alternatively, two or more radial routes may be combined to provide the return journey (Figure 4.4). Link-oriented routes are normally of this kind.

Arc routes join the delivery points in arcs that start and finish at the depot, but which may involve more cross-country travel on poorer roads (Figure 4.5). They tend to be used for local delivery

76 Logistics and the out-bound supply chain

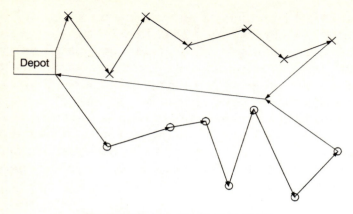

Two radial routes with return to depot

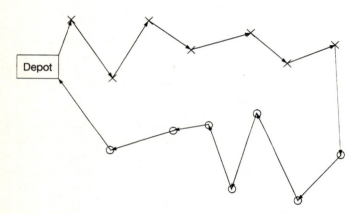

Figure 4.4 *Radial routes*

and collections services, where the delivery and collections points vary from journey to journey.

Area routes cluster geographically close delivery points, but often incorporate a significant dead or non-productive distance at the start and/or the finish of the journey (Figure 4.6). However, they may well be able to make use of motorways or trunk roads to make up time on these stretches.

In determining which customers are to be serviced from which warehouse (if indeed there is more than one), the minimization of delivery costs is a significant factor. Frequently, the use of the nearest warehouse will be the resultant choice but the warehouse giving the shortest travelling time might well be seen as an equally

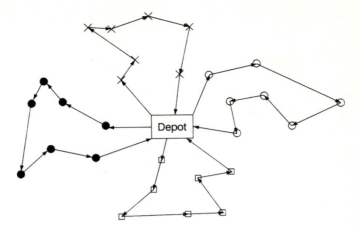

Figure 4.5 *Arc routes*

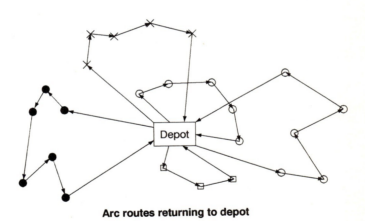

Figure 4.6 *Area routes*

or more satisfactory choice. Other factors that may affect the decision as to which warehouse services a given customer may include product availability and customer preference. Should a customer have been serviced from a location where a good relationship has been established, it may not be sensible to disrupt that situation even if new circumstances make it not the most economically efficient solution.

4.6.2 Selecting a route

Determination of the routes and schedules followed by vehicles may have a number of different (sometimes conflicting) objectives:

- minimizing the distance travelled or fuel cost incurred by the vehicles;
- maximizing customer service levels, however they are defined and which may vary from customer to customer;
- maximizing vehicle utilization and, in particular, driver utilization, which may itself be governed by very strict regulations, and which vary from country to country;
- avoiding the impact of traffic congestion, which is frequently a function of the time of day and the day of the week.

A number of techniques have been developed for vehicle routing, which are discussed below.

The savings method

This technique aims to arrive at a least time (or cost) solution to the problem of allocating a number of different deliveries, which have to made from a single warehouse to some of its customers, to different vehicles journeys. It can be used where the solution is subject to constraints of a maximum vehicle load and also to a maximum journey time. It first determines the journey time (or cost), whichever is the variable to be optimized, from the warehouse to each delivery point i; this is denoted by t_{0i}. The time taken to make the delivery will then be given by

$$t = 2t_{0i} + k_i,$$

where k_i is the turnaround time at delivery point i for unloading, etc. If there are n deliveries to be made, then the total time to make all of them with each delivery made as a separate journey from the warehouse and back again is

$$T = 2(t_{01} + t_{02} + t_{03} + \ldots + t_{0n}) + (k_1 + k_2 + k_3 + \ldots + k_n).$$

Now consider the time taken to go from any delivery point i to any other delivery point j and denote this by t_{ij}. Assuming, as will normally be the case, that

$$t_{0i} + t_{0j} + t_{ij} < 2(t_{0i} + t_{0j}),$$

the saving obtained by visiting i and j in a single combined journey (Figure 4.6) is

$$T_{ij} = t_{0i} + t_{0j} - t_{ij}.$$

A savings matrix $\{T_{ij}\}$ is then compiled of all the possible savings by combining the delivery points in pairs. This matrix is then inspected to determine the greatest saving possible and determine whether it is possible to combine them without contravening any maximum time or maximum load constraints. If it is possible, then this becomes a new route and the procedure is repeated to find further savings. If it is not feasible, the option is discarded and the procedure is repeated to find another potential saving. The process is continued until all possible savings have been taken.

Example

Suppose deliveries have to be made to eight customers, whose travel time from the warehouse and whose delivery loads are given in Table 4.3.

If the travel times between each of the delivery points is given by Table 4.4, the savings matrix $\{T_{ij}\}$ can be determined (Table 4.5).

The total time delivering each load individually to each customer is

$$2(27 + 47 + 64 + 16 + 8 + 78 + 89 + 59) = 776.$$

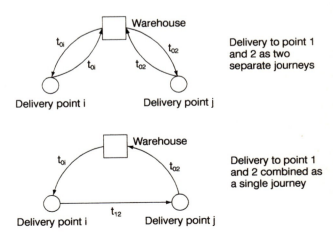

Figure 4.7 *Combining journeys*

Table 4.3 *Customer delivery times and loads*

Customer	Time from warehouse	Load
1	27	34
2	47	52
3	64	17
4	16	48
5	8	23
6	78	11
7	89	59
8	59	42

Table 4.4 *Inter delivery point travel times*

Point	1	2	3	4	5	6	7	8
1	–							
2	33	–						
3	37	64	–					
4	14	30	51	–				
5	34	38	70	20	–			
6	51	78	14	65	84	–		
7	60	87	23	74	93	32	–	
8	31	59	27	45	65	4	22	–

Table 4.5 *Savings matrix*

Point	1	2	3	4	5	6	7	8
1								
2	41							
3	54	47						
4	29	33	29					
5	1	17	2	4				
6	54	47	128	29	2			
7	56	49	130	31	4	135		
8	55	47	96	30	2	133	126	

The largest saving is obtained by combining customers 6 with 7, giving a total load of

$$11 + 59 = 70 \text{ and a time of } 78 + 32 + 89 = 199;$$

so these two delivery points are combined into a single journey and the savings matrix is inspected again; the next largest saving is by combining customers 6 and 8, which gives a load of

$$11 + 59 + 42 = 112;$$

this is in excess of the maximum permitted so this combination is rejected and another attempt is made; the next highest is by combining customers 7 and 3, which gives a load of

$$11 + 59 + 17 = 87 \text{ and a total time of } 78 + 32 + 23 + 64 = 197.$$

On the same basis, customers 8 and 1 can be combined to give a load of 76 and a time of 117, and customers 4 and 2 can be combined to give a load of 100 and a time of 93, leaving customer 5 as a separate delivery. This gives us four journeys with a total time of 423.

Angular ranking method

Another approach to route planning is based on the angles rather than times, and seeks to establish efficient routes within clusters of customers designated to an area type of route. First, the two most distant customers' locations are identified and then customers are chosen according to the least angle formed between the customer and the two most distant locations.

Given a group of orders (Table 4.6) to be delivered to eight customers located according to Figure 4.8, we can see that the two most distant customers are B and F. The least angle is that between A and B; this gives a combined load of 99, so no other customer can be added. The next least angle is between F and G but this leads to a load of 112, which is unacceptable, and so this combination is rejected. The next smallest angle is between F and E, which has a combined load of 81. The procedure is then repeated until all loads are allocated.

Although algorithmic methods for routing are valuable guides to route planning, there are many other factors to be taken into consideration. Travel times frequently vary with the time of day and the day of the week. For some journeys, particularly in cities

Table 4.6 *Customers and loads*

Customer	Load
A	37
B	62
C	15
D	58
E	47
F	34
G	78
H	23
Maximum load =	100

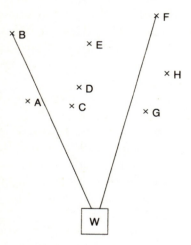

Figure 4.8 *Customer locations in relation to warehouse*

where one-way traffic schemes are in place, the times from A to B are not the same from B to A. The customers may have specific requirements about delivery times to be considered and the drivers themselves can often make minor variations to the route based on their experience and personal preferences. The requirement to pick up fuel may also have to be catered for in planning the route.

Most route schedulers use analytic techniques to prepare a first or second pass schedule, which is then refined manually in conjunction with the driver.

4.7 Summary

Transport is concerned with the physical conveyance of products from source to customer.

The five modes of transport available are: road, rail, water, air, cables and pipelines.

Multimodal transport is when a vehicle of one mode is carried by another mode without unloading the goods (e.g. trucks on ferries).

Intermodal transport is when two different modes of transport are used to convey goods on a single journey.

Packaging is to provide:

- containment of products;
- identification of products;
- protection for products in storage and/or transport.

Primary packaging is that closest to the product, contains the product and provides identification for customers.

Secondary packaging is outside the primary packaging, is largely protective during transport, and provides identification for carriers and regulatory authorities.

Identification increasingly makes use of bar codes.

The transport of goods in bulk from source to a distributor or concentrator, and from a concentrator to distributor or large retailer is the primary transport. It is frequently outsourced to major carriers.

Secondary transport carries smaller loads and delivers direct to customers. Use of own transport facilitates customer contact and provides opportunities for advertising, merchandising and promotion.

In domestic markets, suppliers take most of the responsibility for delivery of their products to their customers. In export markets, a range of different standard terms of trade (Incoterms) are available, which give different responsibilities to the supplier and the customer.

Vehicle routes may be:

- radial – along main roads and then a return back to the warehouse along the same route;
- arc – start and finish at the warehouse with possible cross-country sections;
- area – covering a cluster of delivery points.

Routes may:

- minimize distance travelled or fuel consumed;
- maximize customer service level;
- maximize vehicle or driver utilization;
- avoid impact of traffic congestion.

5

Managing the Supply Chain

5.1 Introduction

In the last three chapters, we have considered the three key elements of logistics and we now need to look at how the total supply chain behaves as a system to supply consumers with goods and services. Most supply chains start at some point with raw materials, either as a result of mining, quarrying and gathering resources, or as a result of agricultural activity. Thereafter, they pass through various stages of processing, production and manufacture before being distributed through supply systems to consumers (Figure 5.1). These industrial activities are frequently classified into three sectors as follows.

The *primary sector*, comprising the exploitation of natural resources:

- mined;
- quarried;
- gathered (fishing, harvesting);
- grown (agriculture).

The *secondary sector*, including a variety of processes:

- extraction;
- refining;
- processing and process manufacture;
- component manufacture;
- sub-assembly and assembly;
- reprocessing;
- construction.

86 Logistics and the out-bound supply chain

The *tertiary sector*, through which they then reach the consumer:

- packaging;
- distribution;
- retail;
- repairs;
- services.

In some instances, notably in the energy industry, for instance oil companies and some water companies, a single enterprise controls the whole supply chain from raw material to end consumer, but this is the exception rather than the rule. It is much more frequently the case that many companies are involved in the total supply chain, and so its management and dynamics are of concern to all companies and organizations that form a part of it.

The supply chain comprises the total network of processes between, at one end, raw materials (extracted or grown) and at the other end, consumers. The extraction and agriculture businesses are at the commencement of the chain, although they are also consumers of equipment, energy, lubricants, fertilizers, etc. The retail industry and the service sector are at the other end of the chain – they directly serve the consumer, but service sector organizations

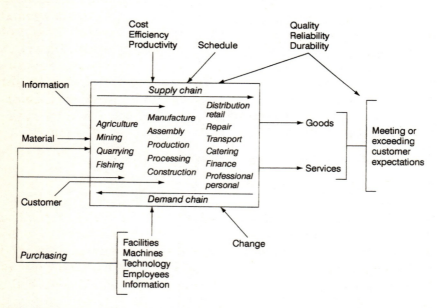

Figure 5.1 *The supply chain and the demand chain*

are also providing their services to the businesses all along the supply chain from raw material processors to retailers. Physical distribution is fundamental to all links in the supply chain to move goods speedily and effectively from one stage to another.

Even what appears to be a relatively simple supply chain may include a large number of organizations. Consider, for example, a set of clothes worn by a typical man or woman, perhaps the clothes you are wearing as you read this book. This might comprise a shirt or blouse, trousers/skirt, jacket or pullover, shoes, socks/stockings/tights and, say, two items of underwear, amounting in total to, say, seven items. These might have been purchased from seven different retailers or perhaps from just one, but typically from about five. Some of these retailers (the larger ones) may have obtained the products direct from manufacturers, but there could have been seven, although more typically three or four, wholesalers involved in the supply chain between the manufacturer and the retailer and inevitably a similar number of transport undertakings carrying the product from manufacturer to wholesaler to retailer. There will, in all probability, have been seven different manufacturers involved in supplying the seven products. Each product is then made up of cloth, thread, fasteners, stiffeners, labels, etc. and each of these items will have further hierarchies of material in their make-up. A typical garment may then have as many as a dozen items in its bill of material so together with distribution, transport and services each item of clothing can easily have 15 elements to its supply chain and the supply chain for a single set of clothes might encompass upwards of 70 enterprises. Furthermore, each enterprise in the supply chain for a given product is likely to be on the supply chains of many other products as well (Figure 5.2).

5.2 The supply chain and the demand chain

Most businesses are embedded somewhere within the chain. The operation of a business is affected not only by its direct customers and suppliers but may also be affected by indirect customers and suppliers, two, three or more steps away in the chain. One way of insulating a business from the performance of other elements in the supply chain, as we have seen in Chapter 3, is to use inventory buffers both upstream and downstream (Figure 5.3).

88 Logistics and the out-bound supply chain

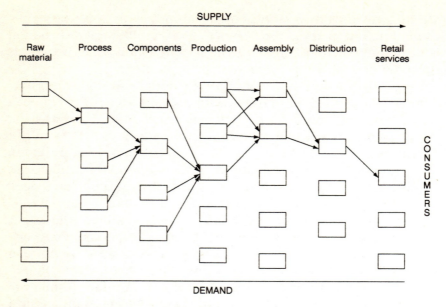

Figure 5.2 *The supply chain and the demand chain*

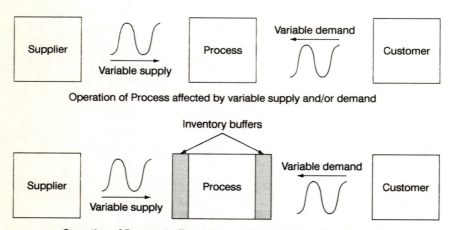

Figure 5.3 *Processes unbuffered and buffered*

This ultimately leads to a supply chain in which all elements hold both upstream and downstream inventory, which as we know from Chapter 3 adds very significantly to the cost of the supply chain (Figure 5.4).

It also has other effects as will be discussed later.

Items in the supply chain are in one of three states:

- *processing*, which is changing the form or condition of the item and is normally adding value;
- *transportation*, in which the item is physically relocated and, if moved nearer to the customer or moved to a place where it is easier to sell, may also be considered to add value;
- *inventory*, which is normally *not* adding value.

Enterprises within the supply chain, in order to minimize their costs, seek to manage their inventories according to the principles discussed in Chapter 3. This implies as accurate a knowledge as is possible of both the demand for products from their customers and the future supply patterns of products from their suppliers. Effective management of the enterprises within a supply chain depend heavily on their ability to have accurate visibility of future potential/actual demand downstream of the supply chain and

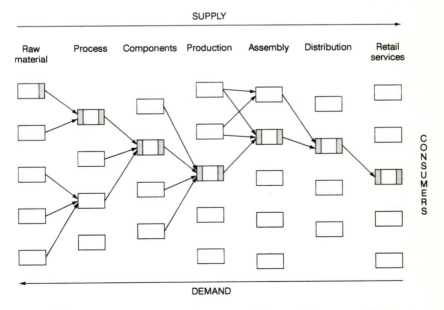

Figure 5.4 *The supply chain and the demand chain with inventory buffers*

accurate visibility of future potential supply upstream of the supply chain. The level of visibility is a function of the order processing systems discussed in Chapter 2; improved visibility may lead to increased order processing costs but lower inventory, so a trade-off exists between order processing and inventory.

Similarly, if the order processing systems provide sound visibility and wide vision of the stock availability, then transport costs can be minimized and a trade-off exists between the order processing and transport systems (Figure 5.5).

Businesses within the supply chain respond to demand generated by the end customers who may be one, two or more steps away in the chain. The demand information is generated through a series of order processing systems passing demand data through the network; this chain of order processing and stock management systems is frequently referred to as the *demand chain*.

So, we can say that:

- through the supply chain flow:
    - goods from raw materials, through various processes to consumers;
- through the demand chain flows:
    - demand information from consumers through various process operators to raw materials suppliers.

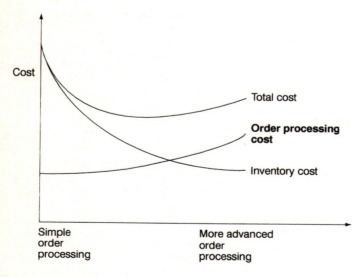

Figure 5.5 *Order processing and inventory trade-off*

In principle, the supply chain should be a reflection of the demand chain with supply exactly matching demand at each stage. In practice, order processing systems, stock control systems, transport scheduling systems, etc., all of which are seeking to optimize some aspect of individual process performance, actually tend to mitigate against this clear reflection of demand in supply.

5.3 Push and pull logistics

We now need to explore the way that the supply chain and the demand chain interact with each other in a manufacturing operation. We shall define two important parameters that are used to assess their functioning.

5.3.1 Customer delivery expectation

This is defined as the maximum elapsed time acceptable to the customer between their placing an order and their eventual receipt of product. If the customer delivery expectation is exceeded, then the customer's perception is one of poor quality of service and this could have a detrimental effect on the business.

The customer delivery expectation is made up of two primary elements: the time taken to transmit the order from the customer to their supplier and the time of the supplier's response, including delivery time (Figure 5.6). Clearly, the shorter the time taken to transmit the order the longer the supplier has to respond; hence the motivation of suppliers to introduce electronic ordering facilities,

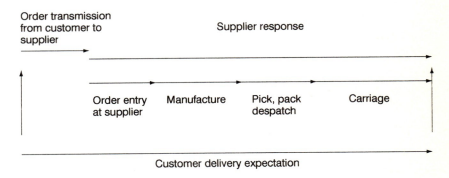

Figure 5.6 *Customer delivery expectation*

which effectively create instantaneous ordering. The supplier's response time is made up of various activities, including order entry, manufacture, picking and packing and delivery. Trade-offs between these activities may have to be made to reduce the overall response time perceived by the customer.

Customer delivery expectations vary from product to product and from customer to customer. In the retail sector, as customers seeking consumer goods, we normally expect to be able to enter a retailer and leave after a few minutes with the products we seek; effectively, in retail, the customer delivery expectation is virtually zero. If we cannot get a product immediately, we leave the shop and go to the competition. For capital goods, however, we may be happy to place an order and either collect or have the goods delivered in a few days. Similarly, in business to business trading, we expect consumables to be available and delivered within, perhaps, 24 hours, but for complex equipment it may be quite acceptable to wait a matter of weeks whilst preparations are made for its use.

5.3.2 Cumulative lead time

This is defined as the elapsed time between the initial commitment of resources (normally the placing of a purchase order) and the availability of the product for supply to the customer (Figure 5.7).

It is determined by analysing the product bill of material and working down from the end product, accumulating the manufacturing and purchasing lead times to determine the earliest purchase that has to be made. So, consider the product A in Figure 5.7. The length of the line from the end product A to its line of sub-assemblies B to G represents the lead time to assemble the product, given the availability of the items B to G; similarly, the line from C to that joining H and I represents the assembly lead time of C, and so on. It can be seen that the earliest item that has to be ordered in order to manufacture A is the component L; hence the sum of lead times for L, H, C, A gives the total time from the placing of the order for L and the product A becoming available for despatch to a customer; this is the cumulative lead time. The cumulative lead times have to include not only manufacturing and procurement lead times but also the time to accept items into stock, initiate manufacturing processes and issue the necessary materials and components.

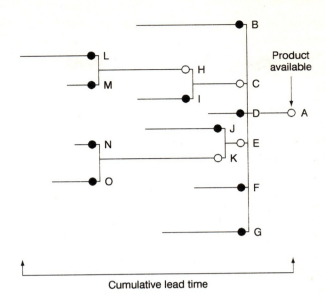

Figure 5.7 *Cumulative lead time*

5.3.3 The fundamental principle of logistics

If delivery expectation is greater than cumulative lead time, then it is possible to undertake all manufacturing after a customer order has been received. All manufacturing and associated purchasing activity can therefore be carried out against a customer order or demand order. Ideally, all orders are completed to arrive 'just in time' to avoid inventory. No commitment of company resources need be made until there is a customer order to cover that resource commitment. This can, in principle, amount to the pursuit of risk-free business. Such systems are known as *pull* systems because work is effectively pulled through the organization by the force of market demand. If, as in the case above, all work can be undertaken against customer orders, we have a pure pull system. This situation may be found in industries such as construction and in other businesses where all work is undertaken under contract from a customer. These businesses do not sell from a catalogue or a product portfolio but sell their skills, resources and competencies at carrying out certain types of work.

If, on the other hand, the customer delivery expectation is so short that no manufacturing or delivery activity can take place within the customer delivery expectation, then the product has to be manufactured and delivered to a place to which the customer comes to buy (e.g. a retailer) before any customer order is received. All this activity has to be undertaken in accordance with a forecast or prediction of what the market will call for. This is referred to as a *push* system. In a pure push system, which is the case with most consumer goods as we have already noted, all activity is to a programme set by the supplier, the programme being a plan based on forecasts, market and product plans, etc.

In practice, most chains and processes are driven by a mixture of push and pull logistics. Those activities that can be accomplished within the customer delivery expectation can be pulled through the organization after a customer order has been received. Those activities that cannot be completed within the customer delivery expectation must be pushed into the system on the basis of a forecast or plan. This includes the purchasing of long lead time items and, frequently, many other purchases and manufacturing processes, depending on the relationship between the cumulative lead time and the customer delivery expectation. By analysing this relationship (Figure 5.10), we can determine which processes lie within the customer delivery expectation and which beyond; hence we can determine which processes can be pulled and which need to be pushed.

Case study

Janssens are producers of high performance plastic components for use in aerospace, automotive and electrical industries. They also supply plastic rod and some plastic material for further working by their customers (semi-finished product). They are part of the Plastic Omnium Group and are based at Maingourmois, near Maintenon, between Rambouillet and Chartres in France. Typical products cost Fr. 700 to manufacture, of which Fr. 100 is labour and overhead cost and Fr. 600 is raw material.

They handle some 400 orders per month; order quantities ranging from small numbers to a few thousand. Order receipt to factory time is typically two days; plant to shipment time is typically two to three weeks. Material is bar coded for identification. Works orders are identified by cards showing order number, customer number, article number, number of pieces, weight, pressure, location of mould lot number, etc. The factory layout is shown in Figure 5.8.

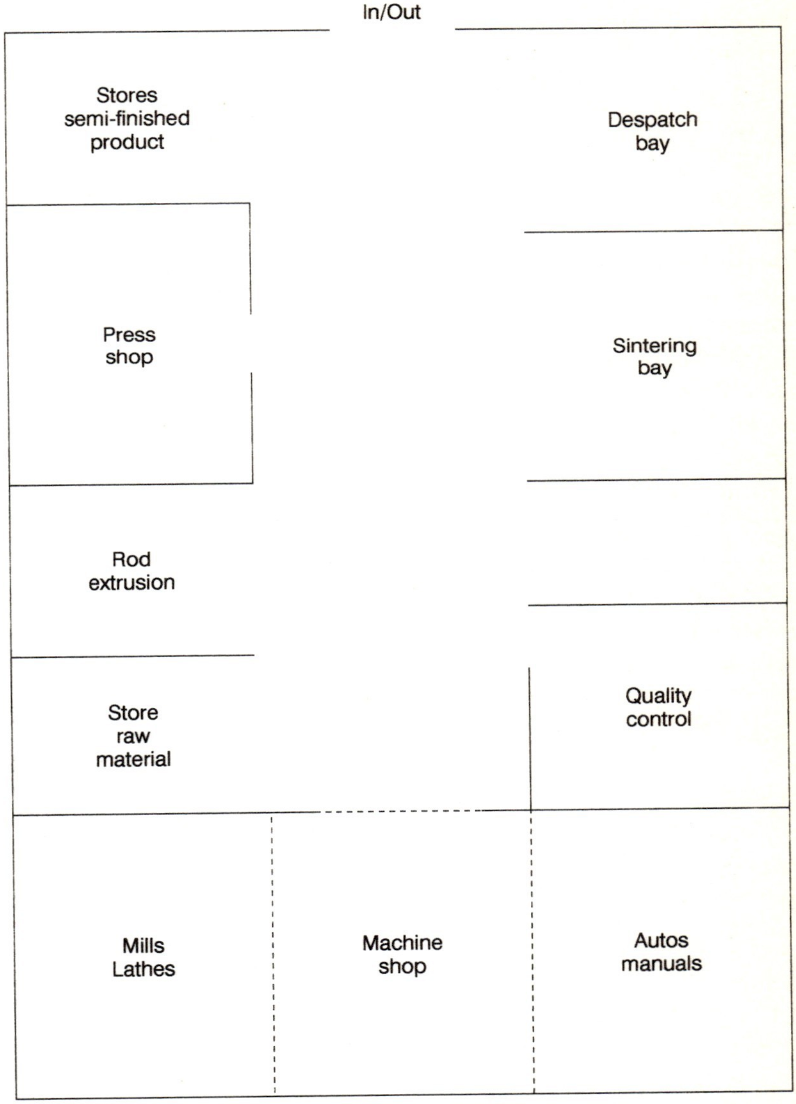

Figure 5.8 *Janssens' factory layout*

Raw material (Teflon and Teflon mixes) is ordered from 10 suppliers as plastic powder or granules and on receipt placed in the raw material store; about two months raw material stock is planned. On receipt of a customer order for a component that is to be produced by the pressing and sintering process, a manufacturing order is generated by the computer system, which authorizes the issue of raw material to the press shop. Here, it is put into moulds and then pressed at pressures of up to 200 kg/cm^2; the pressed components are then passed to the sintering bay where they are sintered at 300°C to produce hardened components. Depending on the requirement from the customer, these components are either despatched at this stage to the customer or passed through to the machine shop for further work. Plastic rods and tubes of various dimensions and cross-sections are produced by an extrusion process operating to the same parameters as the press/sinter process at a rate of between ½ and 1 metre per hour. Because the die changing process in the extrusion facility is quite time-consuming, once production has been initiated for a given specification of rod or tube it is continued for upwards of 6 hours; if this is greater than the customer order requirement, the excess is passed to a stock of semi-finished product in the form of lengths of bar and tube. These may be either sold as 'semi-finished product' or passed through to the machine shop to complete a customer order. Orders received for product produced in this way may therefore be partially or wholly satisfied from stock. Components are then machined to the customer's specifications in a machine shop, including grinders, mills, lathes, drills, etc. They plan to hold 'a little' semi-finished stock in the form of bar and unmachined sections; it is policy not to hold finished product stock. The plant is certified to ISO 9001; the main product quality criteria are physical dimensions, electrical resistivity and heat resistivity. The production process is shown in Figure 5.9.

Those processes to the right and below the wavy line are pulled through the factory; those to the left and above are pushed. The extrusion process undertakes some work that is in direct response to customer orders and some which is undertaken in anticipation of as-yet unreceived customer orders; it is therefore operating to a mixture of push and pull.

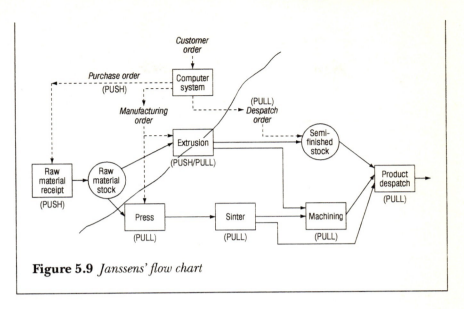

Figure 5.9 *Janssens' flow chart*

The point within a supplier's operation to which a customer's order is uniquely identified is referred to as the *order penetration point*, or OPP (see Figure 5.10). The combination of push and pull is, in fact, found in most businesses and industries. Burger bars, for instance, buy their meat, bread and other food items on a push basis from forecasts of demand, but cook and produce the burgers on a pull basis responding to customer orders. In the automobile industry, the majority of assembly activity and body production is undertaken on a pull basis. Further back in the automobile component industry, the lead times are such that most work is pushed. Many large companies supplying consumer products such as white goods, domestic electronics and automobiles have outsourced all or most activities that are beyond the order penetration point so as to effectively reduce the risk in their business and to avoid the holding of inventory. It is important to note that what appears to be pull for one operator in the supply chain may, in fact, be push. This arises when, for instance, a retailer orders product to replenish inventory on a push basis; their supplier may treat this as a customer order and effectively operate a pull system to satisfy the order, only working when they receive orders from their customer; this is locally pull, but is to the supply chain simply pushing inventory down the chain.

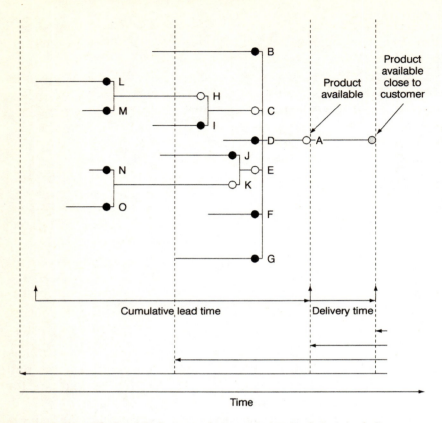

Figure 5.10 *Relationship between the cumulative lead time and the customer delivery expectation*

Within any given company, one tends to find the pull processes at the customer end of the process chain and the push processes at the supplier end.

5.4 Inventory and demand amplification

If there is no inventory in the supply chain at any point, then it will in effect be an accurate reflection of the demand chain. However, in order to meet customer demand patterns, most supply chains include inventory for the reasons we have already explored in Chapter 3. The resulting supply chain no longer accurately reflects the demand chain as it is distorted by the effects of the various stock and transport management systems found in the chain. These have the effect of amplifying the effects of consumer demand variation as we move back up the chain.

Consider a business serving end consumers, which is experiencing genuine consumer demand. If it chooses to hold inventory to provide a service to its customers, it probably carries some safety stock and places orders at intervals, so it also carries primary stock. If the market is rising, then it will be ordering quantities to meet its anticipated demand over the next replenishment cycle, which is a higher amount than that for the current period. This practice continues so long as the market is rising, but as soon as the market changes or slows down, the business cannot respond until, at the earliest, it is due to place its next replenishment order. By this time, they are already experiencing higher than necessary inventory resulting from decreased demand, and so they exaggerate the reduction in their order quantity to overcome this. To their supplier, the fall in demand is both delayed until the order is received and exaggerated, and this tends to continue so long as the market is falling. The effect is further exaggerated by the use of fixed order quantities that are greater than the quantities that would be ordered if the exact order quantity is allowed.

This amplification process is then continued up the supply chain, with increasing levels of distortion from true demand the further we move away from the consumer (Figure 5.11). Effectively, any change in real demand is delayed by the time period of the replenishment

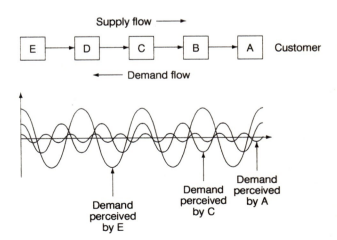

Figure 5.11 *The bullwhip effect*

cycle and may be enhanced both to cover the delay and the constraints of fixed or multiple order quantities. The phenomenon, first identified formally by Forester in 1961, is commonly known as the *bullwhip effect*.

It can be illustrated with the following example. A corner shop trading six days a week and supplying cans of soft drinks to its customers experiences average sales of 25 cans per day; it places an order for replenishment with its wholesaler on Thursdays for delivery on the following Saturday. The wholesaler delivers the cans in boxes of 12, so the corner shop has to order a quantity of 156 cans in the first week although demand is only 150. During the next week, sales begin to climb to an average of 28 cans per day, so finding itself running out of stocks the shop increases its order to 180 cans for the second week. The shop has experienced a 12% increase in sales but the wholesaler has experienced a 15% increase. The next week, the shop experiences a rise in sales to 30 per day (a 7% climb); it responds with an order of 192 cans to its supplier (just less than 7%). Now the following week, the market falls back to 26 cans per day, a fall of 13%, and the shop finds itself with more stock than it needs; it responds by dropping its order quantity to 156, a fall of nearly 19%! If this is repeated all the way up the supply chain, it begins to explain why and how we experience swings in the economy with destocking and restocking of industry.

If all organizations in the supply chain have access to true consumer demand, then this problem can be mitigated, not eliminated, because the fixed order quantities and the non-continuous delivery are also contributory factors. However, giving access to end consumer demand data amounts to providing customer data to one's suppliers and many retailers and other businesses are reluctant to do this. There have been a number of examples in recent years, led by Walmart and Procter and Gamble in the USA, of companies sharing data in this way so that retail consumption is automatically fed to the manufacturer to improve production planning, improve supply to the retailer and mitigate against the bullwhip effect.

5.5 Establishing a domestic distribution system

A number of issues have to be addressed in order to establish the most effective configuration for distribution of a product, trying to balance the perspectives of both the end customer and the parties who make up the supply chain. Many companies, having a (varying) mix of products, need to make judgements that take into account the possible differing requirements of these different products.

First, it is necessary to decide whether goods will be sent direct from source to customer or whether, as is more often the case, they will pass through one or more concentration or distribution points on the way and, if this is the case, how many stages or echelons are required. Then, it is important to determine how many concentration and/or distribution points are needed for each stage in the distribution process, and where they are best located; the means and mode of transport between these locations also need to be determined. These decisions are frequently interdependent, so, in setting the shape and structure of the physical distribution systems, all of these questions need to be considered in concert. The structure and location of the order processing function has also to be established, although this is to an extent an independent issue.

A number of factors will affect these decisions. Firstly, the location of the product source (or sources) and the location of the customers which, between them, define the boundary of the problem. Manufacturers, of course, choose the location of their factories and this choice (which we will discuss in the next chapter) is in part driven by distribution system factors. However, for many products, other factors such as the availability of materials and skilled labour are more significant in setting plant location. The weight and volume of the products affect the cost of transport and this cost in relation to the unit value of the products is another factor influencing the shape of the distribution system. Warehousing accommodation, labour and other associated costs are other factors to be considered in any non-direct distribution system. Finally, and most importantly, the level of customer service sought is a major driver in determining the structure and shape of the physical distribution system.

We have already seen how investment in advanced order processing systems (Chapter 2) can have a beneficial impact on stock-holding, transport costs and customer service. We have also seen how modern technology has enabled the shape of the demand chain (i.e. the network of order processing systems) to become independent of the physical supply chain (warehousing, inventory and transport). There is nevertheless a significant trade-off to be determined between warehousing and inventory, transport and customer service, which we now consider.

As we saw in Chapter 1, performance in a supply chain is a question of a number of criteria but in essence we want to:

- have products available to the customer when they require them;
- supply them at minimum delivery cost;
- make no compromise on quality.

How can we design a distribution system in which the supply chain operates to meet these performance criteria? The holding of plenty of stock close to the customer will clearly provide for product availability, but, assuming that customers are geographically spread, this will lead to stock dispersion and hence greater stock-holding and warehousing costs. Furthermore, stock in the supply chain costs money, as we identified in Chapter 4, and is also a primary cause of the bullwhip distortion. At the other extreme, the holding of all stocks centrally with direct deliveries to customers will lead to lower stock-holding and warehousing costs, but with either a deterioration in customer service or an escalation in transport costs as the opportunity for bulk shipment is lost and the need for high-cost rapid transport to preserve service levels is increased. Nevertheless, these two approaches are increasingly being explored in a number of important industries.

The first, used increasingly for consumer products, locates the majority of the inventory at the consumer end of the supply chain (i.e. at the retailer) and then provides direct consumer demand data upstream to the manufacturer by exploiting the advanced order processing technology now available. Visibility downstream of consumer demand enables upstream operations to work on genuine consumer market data, not data as perceived via a series of stock and transport management systems. Such visibility requires that companies within the supply chain share information. Demand information emanating from retailers' customers is passed directly onto the manufacturers and further upstream as well to their suppliers. In order to manage the stock in the system, the stock data is also shared so that the manufacturer is aware of the stock held at the retailer and the retailer is aware of the stock held by the manufacturer and also the planned production output. In this way, the stock held at the consumer end of the chain comprises a primary cycle stock, which is essentially pulled through the supply chain on a fixed frequency delivery cycle, and a secondary (or safety) stock, which is pushed into the system as a result of planning systems. Upstream of the retailer, the inventory is kept to an absolute minimum so long as the pull systems are in operation. Such systems are commonly referred to as providing *efficient consumer response* or ECR. The sharing of data in this way calls for a high degree of trust between the companies in the supply chain as many companies will inevitably be in more than one product supply chain and competitive supply chains will, from time to time, intersect.

This approach of holding inventory in the supply chain close to the consumer nevertheless implies many stocking points and hence higher safety stocks; it can most easily be justified in the case of high volume fast moving goods. In the example in Figure 5.12, the retailer R1 is a large company owning and running its own distribution system and operating a partnership arrangement with the manufacturer. Customer order data is automatically passed from the point of sale terminals in the shops to the manufacturer and this data has the effect of pulling product through the factory to the retailer's warehouse, which acts as a concentrator, and then on to the shops. Inventory levels in both the warehouse and the shops are kept to an absolute minimum to provide pre-determined customer

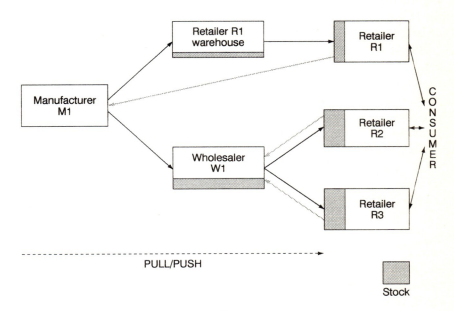

Figure 5.12 *Stock distribution to retailers*

service levels. Retailers R2 and R3 are not large enough to buy direct from the manufacturer so they obtain their product through wholesalers. They place orders when their stock control systems advise them to and place orders with the wholesaler, who then delivers according to a delivery schedule. The wholesaler in turn places orders with manufacturers in a similar way, but, as there is no transmission of genuine customer data through to either the wholesaler or the manufacturer, higher levels of safety stock are maintained and the bullwhip effect is more likely to have its impact.

Where volumes are lower, and so demand patterns more volatile, the levels of safety stock that would be required can become prohibitively high. For such products, like for example automotive spares and many industrial consumables, the stock cannot be held close to the end consumer (except under very special circumstances). Availability of such products is provided by the rapid transmission of orders through the demand chain to a central stock-holding point and then the similarly fast and efficient carriage of the product through to the customer. Again, stock and order visibility up and down the chain is the key to performance and again a high level of trust is called for because most companies are involved in many competing but intersecting chains.

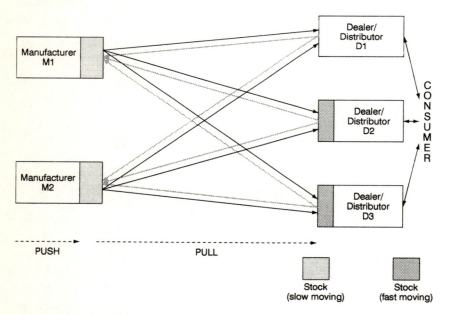

Figure 5.13 *Stock distribution through dealers*

In the example in Figure 5.13, fast moving goods are held in stock in the dealers or distributors. For some dealers, they may be pulled through the system as in retailer R1 above, in which case stocks can be kept to minima to meet customer service levels. For other dealers and distributors, it will be their responsibility to order goods from the manufacturer when they perceive the need and this will probably result in higher stock levels for the same reasons as described above for the wholesaler system. The slow moving products are held centrally at the manufacturer and only ordered by the dealer/distributor when required. This calls for good visibility of product availability from the dealers and also a means of product transport sufficiently fast to meet the customer service levels needed.

5.6 Summary

The supply chain is a network of processes/companies combining to provide end products for consumption.

Most end products result from a complex web of manufacturing and service organizations.

The demand chain is the reverse network of demand information.

In the supply chain, products flow and are either within a process, in transit between processes or as stock.

In the demand chain, information flows through the web of order processing systems.

Customer delivery expectation is the maximum acceptable elapsed time between a customer placing an order and their receiving product.

Cumulative lead time is the elapsed time between the earliest commitment of resource and the availability of product to meet customer demand.

If the customer delivery expectation > cumulative lead time, then pull logistics are possible.

If the customer delivery expectation < cumulative lead time, then push logistics are necessary.

Pull logistics imply working and manufacturing to customer orders.

Push logistics imply working to a plan and manufacturing for stock.

The existence of stock control and transport schemes incorporating minimum order quantities, fixed delivery schedules, safety stocks, etc. can cause amplification of the demand upstream in the supply chain.

For many fast moving consumer goods, stock can be held close to the customer, providing wide availability and their volumes enabling bulk shipment from manufacturers to retailers.

ECR systems feed back consumer demand to manufacture so that production companies can reflect demand and minimize stock in the supply chain.

For slower moving items, more centralized stock systems reduce the total stock-holding but may increase transport costs.

6

International Distribution and e-Business

6.1 Introduction

The final decades of the 20th century witnessed a significant expansion of global trade and the world became a smaller place.

Through radio, television, telephones (mobile and satellite), the Internet, etc., people and companies can communicate with each other, virtually instantaneously, across the world. Similarly, developments in air transport, containerization and the construction of long-distance bridges and tunnels such as Eurotunnel have made the movement of goods over longer distances much more feasible.

At the same time, the World Trade Organization, the EU, NAFTA, etc. have sought to remove the regulatory barriers to the movement of physical goods, workers and funds. The movement of information and 'digital' products has become almost globally deregulated through satellite television and the Internet.

These developments have led in turn to the evolution of both global brands and global products. Manufacturers, competing on both price and quality, seek in turn centralization of production to gain economies of scale and concentration of technical expertise. The challenge of globalization of logistics is the inevitable consequence and the potential options open to a company in getting products from source to customer are numerous (Figure 6.1). The dramatic developments over the past few years of e-business has hastened this logistics challenge and made it more pressing to take up and overcome.

In framing an out-bound logistics strategy, a number of issues have to be faced and questions answered. Should there be a single centralized product source or should production be dispersed? Where should product facilities be located? Should physical distribution be undertaken direct from source to customer or should product be routed through regional, national or local warehouses, wholesalers or retailers? If this is the case, then how many levels of stock or break points – 'Echelons' – should there be in the system? How many stock-holding or break points should there be in each echelon and where should these stock-holdings be located? What means of transport should be employed to ship goods between sources and stock-holdings, between different stock-holdings, and between stock-holdings and customers? What routes should transport take? What activities – warehousing, transport, selling, invoicing, etc. should be undertaken by in-house resources and what should be outsourced? Although some of these have already been addressed in earlier chapters, we now need to attend to the more complex issues that result specifically from the globalization of business and the advent of the Internet and e-business.

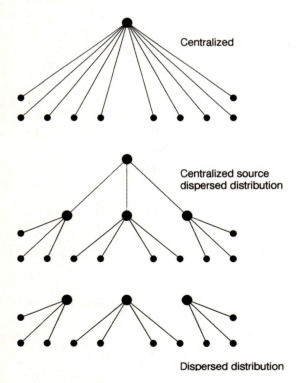

Figure 6.1 *Out-bound logistics options*

The answers to these questions depend on a variety of factors that relate to the nature of the products to be distributed, the markets to which they are to be sent, the needs of the customers and the levels of customer service sought. As it becomes more and more difficult for companies to differentiate on price and quality, logistics parameters such as lead time and availability are increasingly becoming competitive weapons.

6.2 Value density and logistics reach

Cooper (1994) has introduced two concepts that help to provide a framework in which these decisions can be taken; those of *value density* and *logistics reach*.

Value density can be defined in different ways:
either

$$\text{value density} = \text{product value:product weight,}$$

or

$$\text{value density} = \text{product value:product volume,}$$

depending whether product weight or product volume is the determinant of transport cost. Typically, for raw materials and heavy items, it is product weight which is the factor, whilst for many manufactured items that are bulky it is product volume. The value density can also be defined as the following ratio:

$$\text{product value:product transport cost.}$$

Effectively, the value density provides a measure of the transportability of the product. Those with high-value densities, such as high technology equipment or valuable minerals, are much more likely to be conveyed around the world than, say, water or flour.

The *logistics reach* is then defined as:

the maximum economic distance a product can be moved from source to customer.

The logistics reach of a product is clearly a function of its value density but is affected by a number of other factors. If, for instance, we are moving gold bars or diamonds, the need for additional security involved in the transport is likely to increase costs and so reduce the effective logistics reach over other products of the same value density. Items such as fresh food need to be kept chilled and so need special transport facilities to achieve this, which also increases the cost of conveyance. These factors have the effect of reducing logistics reach for a given value density. Other products, such as photocopiers and large computer systems, may require additional services, such as engineering support at installation time, which also has an effect on logistics reach.

Logistics reach tends to be reduced as value density reduces but may be further reduced by either additional transport and logistics costs, or by additional customer service requirements, or both (Figure 6.2).

6.3 Logistics reach and the market

If price is the main driver in the market place for a product, then it is likely that cost will be a key driver in manufacturing and logistics. For products with high-value density, centralized global production will offer economies of scale in manufacture and will therefore be the likely strategy; logistics around this central hub with long distance transport follows, but as the product value is high the higher transport costs can be justified. As value density decreases and the costs of transportation and logistics have a greater operational impact, so production and hence logistics

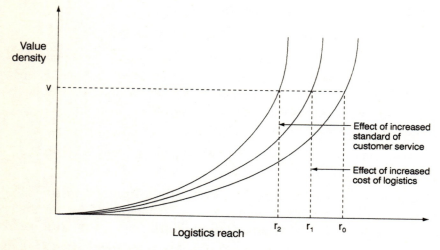

Figure 6.2 *Logistics reach and value density*

become progressively regional, national and local. More production facilities are established, with each providing for its own region, country or local area, so reducing the total cost of transport. Unusually high transport costs associated with raw material, or lack of specialized labour, or other necessary production resources might mitigate against this strategy in exceptional cases. If customer service factors predominate as drivers in the market place, then even for high-value density products we may find that regional production and logistics are appropriate and, as value density falls for this type of product, operations become increasingly localized (Figure 6.3).

Despite the above argument that items with a low-value density have a short logistics reach, there are other products that defy this logic and can travel long distances and still sell successfully. An example of this would be Evian Water, which is bottled at source in the French Alps but is sold successfully in Sydney, Australia. In these situations, factors such as brand reputation and loyalty or product quality become such dominant drivers in the market in place of price that they have the effect of increasing the product value. The customer is prepared to pay for the additional costs of lengthy supply line in addition to the basic product cost even though a cheaper local alternative is available. This has the effect of increasing logistics reach (Figure 6.4).

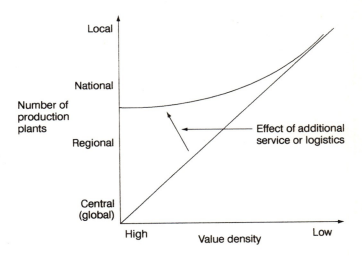

Figure 6.3 *Logistics options with varying value density*

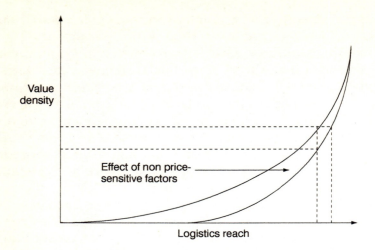

Figure 6.4 *Logistics reach and non price-sensitive forces*

A useful means of expressing the value of a delivery down a supply chain has been proposed by Johansson et al. (1993) in the form:

value = (quality × service level) ÷ (cost × lead time).

If price is the driver in the market place, then reducing cost becomes the operational objective, but this must be achieved without reducing the quality of service level or increasing the lead time; we go for a lean supply chain, centralize as much production and inventory as possible, keep inventory generic and maximize reach. We will refer to this as a *lean* supply chain. In a *lean* supply chain, the order qualifiers are quality, lead time and availability, whilst the order winner is *cost*.

By the same token, if lead times are to be minimized then this must be achieved without increases in cost or deterioration of quality. If service level or availability is the driver, then this must be increased without reducing quality or increasing cost or lead time. We will refer to these supply chains as *agile* supply chains. In an agile supply chain, the order qualifiers are quality, lead time and cost, whilst the order winner is *availability*.

If quality or brand is the sole driver in the market place, then this should normally be achieved without any deterioration in availability or increases in cost or lead time; however, in some cases as we have seen, these may be overridden. We refer to these supply chains as *specific* supply chains. In a specific supply chain, the order qualifiers are availability, lead time and cost, whilst the order winner is *quality* (Figure 6.5).

This approach effectively reverts the argument to the classic business triangle of cost, delivery time and quality. Some authors (Christopher and Towill (2001)) only distinguish between lean and agile supply chains that correspond to the two current manufacturing philosophies. This, however, in the case of logistics, requires a more liberal interpretation of agile than that normally used; i.e. being flexible to provide availability and hence short lead times. This more extended meaning of agile harks back to the Porter approach discussed in Chapter 1. The lean strategy is the applicable logistics approach to a company seeking to achieve cost leadership. For a company seeking focused differentiation, some kind of agility is required but the nature of this agility will depend upon the focused differentiation sought. As yet, no one has successfully found a way of being sufficiently agile to be able to regularly and quickly jump from one focus to another and at the same time remain focused!

6.4 Distribution and inventory strategies

Discounting the specialized supply chain where issues such as product quality and/or functionality are the main drivers in the market, and where both the cost and the delivery performance in terms of either lead time or availability are low-level qualifiers, the key choice of logistics is between the lean and the available/agile.

	Qualifiers	Winners
Specialized supply	Availability Lead time Cost	Quality
Agile supply	Quality Lead time Cost	Availability
Lean supply	Quality Lead time Availability	Cost

Figure 6.5 *Supply characteristics*

The lean supply chain, as we have already noted, will favour centralized production, which may be either global for high logistics reach or centralized to a region or locality for lower reach items. The lean supply chain will also favour the holding of the majority of inventory as raw material rather than finished product because this keeps open the options of what product will be made with the material until the customer order is received. This in turn means fewer stock held items and stock being held without any added value content included; hence lower stock carrying costs. By holding stock as raw material and also centralizing production, delivery lead times are extended and so the time needed to respond to customer orders is increased. This means reducing order receipt and processing times to as near to instantaneous as possible and reducing the manufacturing lead times or persuading the customer to extend their delivery expectation. If price is genuinely the driver in the market the latter may be acceptable, but as lead time remains a likely order qualifier, even in lean supply chains, the options for this approach may be limited. Internet-based order processing may reduce this time but the scope for reductions is limited by the very short times already in operation. This leaves the reduction of manufacturing lead times as the only other option in this situation.

The agile supply chain, on the other hand, will favour holding inventory as finished product stock as close to the customer as possible so as to provide both a high level of availability and short lead times. For items with a long logistics reach, a centralized manufacturing facility is still an option, with the finished product stock then held in a multiplicity of warehouses around the world and the penalty of high inventory carrying costs that may be borne in an agile environment where cost is but a qualifier. Short delivery expectations can be met at a cost. For items with a short logistics reach the establishment of a multiplicity of manufacturing locations is the favoured option because it both overcomes the problems of reach and facilitates availability.

This approach to the agile supply chain, however, has a feature that appears to be the antithesis of agility. By holding most of the stock as finished product the manufacturing process has been committed; it has lost the agility to revert to another product to respond to a market change that it would retain if it held as much inventory as possible in either raw material or part finished product sufficiently generic to leave the option open. Hence a means of obtaining a balance between these two scenarios is sought.

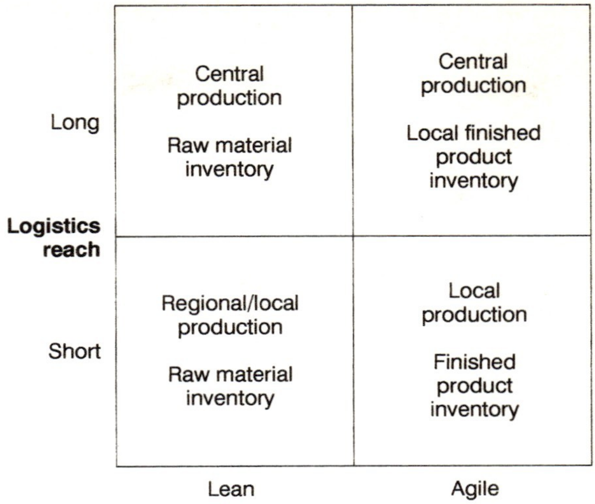

	Lean	Agile
Long	Central production Raw material inventory	Central production Local finished product inventory
Short	Regional/local production Raw material inventory	Local production Finished product inventory

Logistics reach (vertical axis)

Figure 6.6 *Inventory and production strategies*

We need to identify how far real demand penetrates the supply chain, i.e. where market pull meets upstream push; we call this the *decoupling point* (Figure 6.7).

We attempt to postpone until the last possible moment the point at which inventory is identified to a particular demand, i.e. we seek to keep generic inventory as close to the market as possible whilst maximizing product availability.

The first question is the market extent of the product. If a product is global in all respects, then a logistics strategy has to be developed to handle its manufacture and distribution to cover the full extent of its market. If the product has a more restricted market, then the strategy has only to fulfil the needs of that domain. Manufacture within that geographic market will be the natural conclusion from the standpoint of distribution. If some other manufacturing criterion, such as labour of raw material availability, leads to a contrary conclusion then the distribution cost from source to market will be factored into any decision to manufacture outside the market area.

For a genuinely global product, there are a number of possibilities.

If the logistics reach is long and the customer delivery expectation is also long, then the centralization of production to a single source for distribution to global markets, creating a lean supply chain, is perfectly feasible.

Logistics and the out-bound supply chain

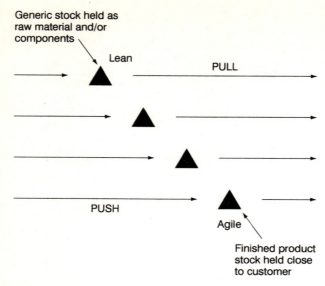

Figure 6.7 *Decoupling point and inventory location*

If the delivery expectation is shorter and the requirement from the supply chain becomes more agile, then how can greater availability and shorter lead times be achieved?

1. The product is genuinely global without any regional variations in either its final make-up or in packaging. Then there may be sufficient volume and minimal risk in holding some finished product stock centrally without losing too much flexibility in the use of raw materials or components for alternative products.
2. The product has regional or local variations in either its final make-up or its packaging. Then a centralized production facility and holding inventory up to the point at which the product is generic is practicable. This is followed by distribution to regional or national centres, where the variants are added on a pull basis to meet the local market demand.
3. The product has a genuine global make-up and market but has a shorter logistics reach than can justify a centralized production capability, or the customer delivery expectation is less than would be satisfied by a global source of product supply. The product may contain key components which, as components, have much longer logistics reach than the finished product

itself. In this case, these components can be manufactured in a single global source facility and then distributed to regional or local assembly plants that hold inventory at this level. This overcomes both the reach problem and facilitates more agility in supply to satisfy the shorter delivery expectations. Figure 6.8 summarizes the manufacturing and supply options available.

6.5 Facilities location

The location of plant is also critical in establishing an international distribution system. When local plants are needed, the driver is normally delivery lead time to customers and so customer proximity is an important factor. Access to transport and communications infrastructure will also be significant determinants in selecting an appropriate location. The products supplied from these plants are characterized by low logistics reach, normally resulting from low-value density and therefore are low technology

Configuration	Provides
Centralized production Raw material inventory	Lean production, lean supply, flexibility, long lead times
Centralized production Central inventory of finished product	Lean production, lean supply, less flexibility, long lead times, more inventory, shorter lead times
Centralized production	Lean production, cost of finished product stock
Regional/local inventory of finished product	More agile supply, shorter lead times, availability
Centralized production to generic level	Lean production to generic, cost of generic stock
Regional/local assembly & inventory of finished product	More agile supply, shorter lead times, availability
Centralized component production	Lean component production, cost of component stock
Regional/local assembly & inventory of finished product	More agile supply, shorter lead times, availability
Regional/local production Regional/local inventory of finished product	Agile production and supply

Figure 6.8 *Manufacturing and supply options*

items; hence availability of skilled labour is not normally an issue. Should, however, the logistics reach have been lowered by the need for customer support in commissioning and installation of equipment, then the availability of these resources may be a factor in plant location.

If a regional or an area plant is being considered to serve the market in a number of neighbouring countries, then the import regulations in respect of any key components together with financial incentives (grants), tax regimes, and political stability all become factors to be considered in selecting the location for a manufacturing facility. In this context, locating a manufacturing and/or distribution facility within a trade block such as the EU or NAFTA or within a currency area such as the Euro will have clear advantages over being outside; the extent to which these advantages may or may not be off-set by tax or other regulatory regimes has to be investigated.

When a global facility is to be established, or even a regional one supplying a significant global region then, because there is only a single, or at the most a very few, plants from which product can be supplied, the location selected is more critical. Because the output from these plants is being supplied on a lean rather than an agile basis, the production cost and therefore production-related factors are more important criteria than those of distribution. Availability of key skills, know-how, technology and materials are the driving factors predominating over distribution-related criteria in determining location. Export regulations are, however, more significant than import. Tax regimes, financial incentives and political stability are also key factors to be considered.

6.6 The impact of the Internet

Over the past four or five years, the rapid growth of the Internet has made massive inroads into the business world and this is having a significant bearing on distribution systems. The four key interactions for a supplier interacting with its market are shown in Figure 6.9. Three of these are:

1. Communication and promotion of products, prices, etc. from supplier to customer.
2. Communication of orders from customer to supplier.

3. Issue of invoices from supplier to customer and payment of the invoice.

These are all capable of being undertaken entirely on the Internet and we are witnessing a major shift of business in this direction. In the first few years, the Internet was used largely as a marketing tool as in (1) above; companies presented their product portfolio on their web pages, used it as an advertising medium and to distribute their catalogues. Customers then placed orders using traditional methods; more recently, the Internet has become a mechanism for order entry, invoicing and payment (2), (3). Many companies now offer customers a discount for ordering on the net as the customer is fulfilling the function of order entry and saving the supplier these costs. Once the order has been entered, it can be priced automatically, the customer can establish creditworthiness, the order can be priced and invoiced; through the use of credit cards, Internet banking, electronic funds transfer and other Internet-related mechanisms, payment can also be effected with minimum involvement from the supplier's personnel. This facility makes it potentially possible for every company to sell its products into global markets – *if they can deliver them*!

This then is the distribution challenge raised by the Internet (Gurau et al (2001)). Some of the earlier companies into Internet trading failed on just this issue. In a study by Arthur Andersen in 2000, reported by van Hoek (2001), logistics was reported as a major contributor to five of the top eight problems reported by customers trying to use the Internet as a means of procurement. Other than products convertible to digital format (i.e. text, graphics, music,

Figure 6.9 *Supply and demand chain transactions*

information, etc.), physical distribution has to be accomplished by traditional means. The strategies for distribution have to be established and followed but Internet trading is raising the level of expectation of the customer, whether end-consumer (B2C) or commercial (B2B). By the use of bar coding, tracking systems of product distribution, in addition to making information available to the supplier and manufacturer, can now be used to provide on-line tracking to the customer so reducing the number of customer queries arising.

Competitive strategies have gone full circle over the past 50 years. In the years immediately following World War II, product availability was the key to commercial success. Because of the shortages in manpower, materials and machinery resulting from the war, almost any product that a manufacturer could produce could be sold. Once these shortages had been overcome, it was productivity and efficiency that became the competitive weapons. By the 1970s, cost control had been widely achieved and, through Japan, quality became the competitive criterion by which companies were judged. Today, most companies have found their way through the quality issue and, for those not seeking cost leadership, the most competitive differentiator is becoming delivery – delivery speed, delivery reliability and punctuality – availability all over again but at a much greater level and with quality and cost efficiency thrown in as strict order qualifiers. The Internet has further accentuated this; delivery promises once quoted in months and weeks are now specified in hours and even minutes.

6.7 Summary

Global trade is experiencing massive growth as a result of both technological advances in communications and transport systems and also following political developments establishing trade blocks, currency blocks and the relaxation of trading regulations.

Distribution options range from highly centralized to very dispersed.

Value density is a measure of the relationship between product value and unit transport cost.

Logistics reach is the maximum economic distance a product can travel source to customer.

Logistics reach is reduced by additional logistics or customer service requirements.

Logistics reach may be increased by brand reputation or product quality.

Distribution value down a supply chain may be measured as:

(quality × service level) ÷ (cost × lead time).

Lean supply chains seek to minimize cost and waste; their order winner is cost.

Agile supply chains seek to offer response and delivery performance; their order winner is availability.

Different production and inventory strategies result from the combinations of lean and agile objectives: long and short logistics reaches.

The decoupling point is the point in the supply chain where the upstream push meets the downstream pull.

Postponing the decision to use stock to meet a specific product or market need until as far down the supply chain as possible increases agility.

The number of production facilities depends on logistics reach: few for long-reach products; many for short-reach items.

If there are many plants, their location will be a function of customer location and local logistics.

If there is only one (or few) plant(s), manufacturing factors will determine location together with international distribution factors.

The Internet is accentuating the logistics challenge by increasing the potential for international business.

The Internet can enhance order processing, including invoicing and payment systems. It can only deliver digital products.

The Internet is raising customer expectations of delivery but for most products they still have to be delivered by conventional means.

Appendix: Typical Examination Questions

Chapter 1

1. What are the performance measures for systems distributing:
   a cosmetics;
   b milk;
   c automobiles?
2. Why do the measures for the above products differ?
3. How does inventory management interact with transport in the distribution of:
   a domestic gas;
   b garden furniture?
4. How would one assess delivery reliability for the delivery of water? How does it differ from that of fashion clothing?
5. Identify two products for which delivery speed is the priority and two for which delivery reliability is the priority. Contrast their distribution systems.
6. When might order reliability be an order winner? Give an example.

Chapter 2

7. Given that sale of garden sprinklers over recent quarters is:

Quarter	Sales of sprinklers
1	2621
2	2593
3	3408
4	2598
5	2725
6	3643
7	3527
8	2713
9	2457
10	3746
11	3865
12	2741
13	2529
14	3875
15	4026
16	2781
17	2654
18	3954
19	4216
20	2956

   determine the centred moving average.
8. For the sales figures in the previous table, by calculating the seasonal factors determine the trend.
9. If demand rather than sales data had been available what differences might be expected?
10. Why is order tracking important in distribution?
11. In what situation would judgemental forecasting be appropriate? Give an example.
12. How has the combination of telecommunications and computing technology affected the way goods are ordered by a small retailer?

Chapter 3

13. What are the main purposes of holding stock?
14. Why can free stock become negative? What actions does it suggest?
15. Compare and contrast the use of average and standard costing systems.
16. What are the advantages and disadvantages of fixed frequency ordering systems?
17. Stock holding costs are 10% of unit cost per annum, ordering costs are £30 per order and usage of a given item is 6000 per annum. It can be purchased for £40 per unit or for £36 per unit if the order quantity is greater than 300. How much should be ordered?
18. Why does the quantity of safety stock increase as the number of stock holding warehouses increases?

Chapter 4

19. Compare and contrast the use of road and rail transport for the carriage of food. What sort of food would you recommend for each mode?
20. What are the roles of packaging? What are the uses of identification labels?
21. Distinguish between a concentrator and a distributor. How does primary transport differ from secondary?
22. What are the advantages and disadvantages of undertaking one's own transport operations?
23. Compare and contrast selling a product on an FAS basis with selling on a DDP basis.
24. Compare and contrast radial and arc routes. Describe situations where each would be appropriate.

Chapter 5

25. How many enterprises contributed in the supply chain that provided your dinner last night? Try to map the supply chain.
26. What is the demand chain? How does it relate to the supply chain?
27. What does inventory do in the supply chain? How does it help? What are the downsides?
28. Compare and contrast push and pull logistics.
29. What is the bullwhip effect? What are its primary causes?
30. Compare and contrast holding stock close to the customer with holding it centrally. Give examples where each might be appropriate.

Chapter 6

31. How can logistics reach be affected by factors other than value density?
32. What are the likely company strategies that would call for (a) a lean, and (b) an agile supply chain? Why?
33. What is the significance of the decoupling point?
34. What are the advantages of global centralization of the production of components whilst dispersing assembly facilities? For what products might this be done?
35. What would be the characteristics of a country that would be important in deciding whether or not to establish a manufacturing facility?
36. How might the Internet change distribution systems in coming years?

Bibliography

Christopher M, *Logistics and Supply Chain Management*, published by Pitman, 1992.

Christopher M and Towill D, 'An Integrated Model for the Design of Agile Supply Chains', *International Journal of Physical Distribution and Logistics Management*, vol 31, no 4, pp.235–254, 2001.

Cooper J, 'Logistics Strategies for Global Businesses', *International Journal of Physical Distribution and Logistics Management*, vol 23, no 4, pp.12–23, 1993.

Cooper J (ed.), *Logistics and Distribution Planning* (2nd edition), published by Kogan Page, 1994.

Fawcett P, McLeish R and Ogden I, *Logistics Management*, published by Pitman, 1992.

Forrester J W, *Industrial Dynamics*, published by MIT Press, 1961.

Gourdin K N, *Global Logistics Management*, published by Blackwell, 2001.

Gurau C, Ranchod A and Hackney R, 'Internet Transactions and Physical Logistics: Conflict or Complementary?' *Logistics Information Management*, vol 14, no 1/2, pp.33–43, 2001.

Hill T, *Manufacturing Strategy*, published by Macmillan, 1993.

van Hoek R, 'E-Supply Chains – Virtually Non-existing' *Supply Chain Management*, vol 6, no 1, pp.21–28, 2001.

Johansson H J, McHugh P, Pendlebury A J and Wheeler W A, *Business Process Re-engineering: Breakthrough Strategies for Market Dominance*, published by John Wiley and Sons, 1993.

Porter M E, *Competitive Advantage: Creating and Sustaining Performance*, published by Free Press, 1985.

Background and Rationale of the Series

This new series has been produced to meet the new and changing needs of students and staff in the Higher Education sector caused by firstly, the introduction of 15 week semester modules and, secondly, the need for students to pay fees.

With the introduction of semesters, the 'focus' has shifted to module examinations rather than end of year examinations. Typically, within each semester a student takes six modules. Each module is self-contained and is examined/assessed such that on completion a student is awarded 10 credits. This results in 60 credits per semester, 120 credits per year (or level to use the new parlance) and 360 credits per honours degree. Each module is timetabled for three hours per week. Each semester module consists of 12 teaching weeks, one revision week and two examination weeks. Thus, students concentrate on the 12 weeks and adopt a compartmentalized approach to studying.

Students are now registered on modules and have to pay for their degree per module. Most now work to make ends meet and many end up with a degree and debts. They are 'poor' and unwilling to pay £50 for a module textbook when only a third or half of it is relevant.

These two things mean that the average student is no longer willing or able to buy traditional academic text books which are often written more for the ego of the writer than the needs of students. This series of books addresses these issues. Each book in the series is short, affordable and directly related to a 12 week teaching module. So modular material will be presented in an accessible and relevant manner. Typical examination questions will also be included, which will assist staff and students.

However, there is another objective to this book series. Because the material presented in each book represents the state-of-the-art practice, it will also be of interest to professional engineers in industry and specialist practitioners. So the books can be used by engineers as a first source reference that can lead onto more detailed publications.

Therefore, each book is not only the equivalent of a set of lecture notes but is also a resource that can sit on a shelf to be referred to in the distant future.

Index

air transport 64
amplification, inventory and demand 98
angular ranking method 81
arc routes 75
area routes 76
associative prediction 31
average cost 47
averages
  costing 47
  moving 27, 28
    centred 28

buffered, processes 39, 88
  unbuffered and 39, 88
buffers, inventory 89
bullwhip effect, the 99

cables transport, pipelines and 64
call centre, centralized order processing at a 34
carriage of goods, transport system and 14
centred moving averages 28
communication 13
  information technology and 13

competitive advantage 9, 11
  strategies of 11
concentrators, transport via distributors and 72
containment 66
cost 9
  actual 45
  average 47
costing 47, 48
  average 47
  standard, systems, 48
C-terms 73
cumulative lead time 92, 93
customer 7
  delivery expectation 91
  service level 7, 15
cycle stock, primary or, 48
cyclical demand 25

dealers, stock distribution through 104
decoupling point 115
delivered duty paid (DDP) 73
delivery expectation, customer 91
  reliability 5
  time, total 4

Delphi group forecasting   24
demand   25, 87, 98
  amplification, inventory and
      98
  chain   87
    supply chain, and   87
  cyclical   25
  de-seasonalized demand   30
  seasonal   25
discounts, effect of,   54
distribution
  domestic system   100
  international, e-business and
      107 et seq
  stock   103, 104
    dealers, through   104
    retailers, to   103
  strategy   9, 113
    inventory and   113
  systems   3, 12
    product   3
distributors, transport via   72
  concentrators and   72
domestic distribution system   100
D-terms   73

e-business, international
    distribution and   107
economic order quantity   52
efficient consumer response (ECR)
    103
electronic data interchange (EDI)
    34, 35
  systems   35
EOQ   61
E-terms   73
expert opinion   23
EXW (ex-works)   73

facilities location   117
FAS   73

forecasting   17
  Delphi group   24
  judgemental   23
  order management and   17 et
      seq
  statistical   24
free stock   61
F-terms   73

identification   66
in-bound supply chains   3
information technology   13
  communication and   13
inland water transport, sea and
    64
intermodal transport, multimodal
    and   65
international distribution,
    e-business and   107 et seq
Internet, impact of the   118
inventory
  accounting for   45
  amplification, demand and   98
  buffers   89
  strategies, distribution and   113
  trade-off   90
    order processing, and   90

judgemental forecasting   23

life cycle of an order   22
logistics
  fundamental principle of   93
  options, out-bound   108
  push and pull   91
  reach   109, 110
    the market, and   110
    value density and   109

manufacturing options, supply and
    117

market logistics reach, and the 110
  surveys 23
moving averages 27, 28
  centred 28
multimodal transport, intermodal and 65
multiple warehousing 59

order
  completeness 6
  life cycle of an 22
  management, forecasting and 17 et seq
  penetration 97
  processing
    centralized 34
      call centre, at a 34
      inventory trade-off, and 90
      stock records, and, 40
      systems 31
  qualifiers 11
  quantity, economic 52
  trigger 58
  winners 11
out-bound logistics options 108
  supply chains 3
ownership, responsibility and 70

packaging
  primary 67
  secondary 67
pipelines, cables and 64
primary packaging 67
  sector 85
  stock, control of 49
    cycle, or 48
  transport, or trunk 69
process chain 39, 88
  buffered 39, 88
  unbuffered 39, 88

processing 89
product availability 8
  distribution system 3
protection 66
push and pull logistics 91

radial routes 75
rail transport 63
ranking method, angular 81
reorder point 51
replenishment trigger 58
responsibility and ownership 70
retailers, stock distribution to 103
road transport 62
routes 75, 76
  arc 75
  area 76
  radial 75
  types of 75
routing 75
  vehicle 75

safety stock 48, 56, 61
  control of 56
  secondary, or 48
scheduling, vehicle 75
sea transport, inland water and 64
seasonal demand 25
  factors 29
secondary
  packaging 67
  sector 85
  stock, or safety 48
  transport 69
standard costing systems 48
statistical forecasting 24
stock allocation 20
  control 48
  cycle, primary or 48

distribution   103, 104
   dealers, through   104
   retailers, to   103
free   61
handling, storage and   42
management   13
   cycle, or   48
   primary, control of   49
records   40, 61
   order processing and   40
safety   48, 56, 61
   control of   56
   secondary, of   48
secondary, or safety   48
storage, stock handling and   42
supply
   chains   3, 4, 85 et seq, 87, 117
      demand and   87
      in-bound   3
      managing the   85 et seq
      out-bound   3
   options, manufacturing and   117

time series projections   25
total delivery time   4
trade-off, inventory
   order processing and   90

transport   62 et seq
   decision, the   68
   primary, trunk or   69
   secondary   69
   systems, carriage of goods and   14
   trunk, primary or   69
   via concentrators   72
   via distributors   72
   via wholesalers   71
transportation   89
trigger   58
   order   58
   replenishment   58
trunk transport, primary or   69

unbuffered, processes   39, 88
   buffered, and   39, 88

value density, logistics reach and   109
vehicle routing   75
   scheduling   75

warehousing, multiple   59
   operations   44
wholesaler, transport via   71
World Trade Organization   107